BIOLOGY AND EFFECTS OF ENDOSULFAN AND DIMETHOATE ON MARINE MOLLUSCS

BIOLOGY AND EFFECTS
OF ENDOSULFAN AND DIMETHOATE ON MARINE MOLLUSCS

DR. (MRS.) T. SARADHA

NIDHI BOOK CENTRE
DELHI (INDIA)

First Pulished : 2004

Published by
NIDHI BOOK CENTRE
3421-A, IInd Floor, Narang Colony, Tri Nagar,
Delhi-110035. Phone : 27153083, 30967505 (O) , 27522129 (R)

ISBN : 81-902086-1-6

Laser Type Setting by
A-ONE GRAPHICS
X-4, Street No. 2, Brahmapuri, Delhi-110053
Phone : 2218 3470

Printed at
Salasar Imaging Systems
Delhi, Ph. 31045804

FOREWORD

I am extremely happy to go through the investigations made by **Dr. (Mrs.) T. SARADHA** to understand the ecology, breeding biology, the bio-chemical composition of the Tellinids *T. ala* and *T. cuspis*. This study has a significant role to play on Indian Economy, for the Tellinids are used by the coastal people for consumption.

The people of this area must know the effect of two pesticides endosulfan and dimethoate on the histology of Tellinids. She has analysed the various aspects of its culture and the methods of its growth and development, the nutritive value particularly of the samples from Porto Novo waters. This will enable the common folk to culture them by inexpensive methods.

Dr. T. Saradha's conclusion that salinity has an effect on the reproduction gives ample scope to further up studies. Besides tracing the biochemical influence on the Tellinids, she has proved how the rate of filtration decreased with an increase in the test concentration of pesticides. Interestingly, the histopathological studies have shown how the bodies of *T. ala* are affected by endosulfan concentrate. She has crisply emphasized the fact the knowledge of Tellinids, the spawning periods, the breeding methods, the physio-chemical norms and the nutritive value will go a longway to adopt the successful methods for culture practices.

Since pesticides are causing great concern, people need to take suitable measures for the replacement of such dangerous pesticides so as to develop aquaculture. Dr. Saradha's investigation need to be continued further by other scientists to promote the Indian economy. Her thesis is a model to be emulated by others in the sense that it is a painstaking and devoted work. She richly deserves to be complimented for her outstanding work which is the need of the hour.

12.12.03 **Dr. M.D. Sampath**
Mysore

PREFACE

This is an exhaustive project work on the biology and effects of pesticides endosulfan and dimethoate on the physiology of Tellinids (Marine bivalves).

An extensive field work on the coastal tract has been undertaken to collect the data. Keeping in view the recent trends on various aspects of biology, breeding habits which is essential for culture production of bivalves, it is a dessideratum to examine the various aspects of the recent trends in this field.

In these days of dwindling reserves of natural resources, an approach has to be made to understand, preserve and conserve our natural bio-treasures. Exposure to various pollutants have wiped away most of our marine, fresh water, terrestrial bio-wealth. It is the moral and academic responsibility of the academicians of today to impress upon the students the importance of bio-diversity and their conservation.

It is with great pleasure I express my sincere thanks to Drs. A.D. Ansell, Dustaffnage, Marine Research Laboratory, Scotland, Rosspohlo, Californnia State University, Guy Bachelet, France and Talbot, South Africa for the voluminious supply of literature and constructive criticisms that inspired me in the present work.

I wish to express my deep sense of gratitude and thanks to my beloved parents, brothers, sister, my husband and my daughter for their understanding, willing, co-operation and constant encouragement to complete my work successfully.

I am highly indebted to Dr. M.D. Sampath, Retd. Director, Archaeological Survey of India, Mysore who has seen the pages through press and to the forword given by him.

Finally I wish to place on record my sincere and grateful thanks to Sri C. P. Gautam, Bharatiya Kala Prakashan, New Delhi, for having come forward to publish my thesis and to print the same in record time.

16-12-2003 **T. SARADHA**

ACKNOWLEDGEMENTS

It is with pleasure I express my sincere gratitude to Dr. R. KASINATHAN, Junior Research Officer, CAS in Marine Biology, Annamalai University for suggesting the problem, supervision and guidance throughout the course of the study.

I am also immensely grateful to the Director, CAS in Marine Biology for giving facilities for conducting this research work in Marine Biological station, Porto Novo.

My thanks are due to Drs. A.D. Ansell, Dunstaffnage Marine Research Laboratory, Scotland, Rosspohlo, California State University, Guy Bachelet, France and Talbot, South Africa for voluminous supply of literature and constructive criticisms that inspired me in the present investigation.

I am also thankful to all staff members and Research Scholars of CAS in Marine Biology for their constant encouragement during the period of study.

The Indian Council of Agricultural Research (ICAR) is gratefully acknowledged for providing Research Fellowship during the tenure of this research work.

I wish to express my deep sense of gratitude and thanks to my beloved parents, brothers, sister, my husband and my daughter for their understanding, willing co-operation and constant encouragement to complete my studies successfully.

(T. SARADHA)

CONTENTS

LIST OF FIGURES

LIST OF COLOUR PLATES

1

GENERAL INTRODUCTION

1.1. PREFACE

Bivalvia is a large and highly successful class of molluscs which includes forms such as oysters, mussels, clams, cockles and scallops. These have been used as sea food from early times by man, since they are available in plenty and could be easily collected from rocks, sandy and muddy intertidal areas of the seas and estuaries. Molluscs occur abundantly in the Vellar estuary, Southeast coast of India. Mussels and oysters are plentiful on rocky coasts. The sandy and muddy sea shores and estuaries have very extensive beds of various kinds of clams. The clams occurring commonly in Indian waters are *Meretrix* sp., *Katelysia opima* (Gmelin), *Solen kempi* Preston, *Paphia* sp. and *Donax* sp. (Durve, 1964). Many investigations were carried out on the biology, biochemistry, physiology and various aspects of the class Bivalvia. Noteworthy among them are as follows: studies on the anatomy and physiology of *Sanguinaria diphos* were carried out by Venugopalan (1955). Kasinathan (1963) studied the Alloxan induced 'diabetes[1] in the estuarine clam *Meretrix casta.* Kasinathan and Vijayaraghavan (1964) reported the effect of insulin on the amino nitrogen content of free amino acids in the blood and foot of *Meretrix casta.* Venugopalan (1955) studied the fatty acid composition of the fat in male and female oyster, *Crassostrea madrasensis.* Krishnamurthy (1964) investigated the distribution of iron in the tissues of some bivalve molluscs occurring in Vellar estuary. The biology of the venerid clam *Meretrix casta* (Balasubramaniyan, 1984), M. *meretrix* (Jayabal, 1984) *Anadara rhombea* (George John, 1980), *Katelysia opima* (Kalyanasundaram, 1982), *Sunetta scripta* (Diravidamani, 1985) and *Crassostrea madrasensis* (Marimuthu, 1986) were carried out recently.

The family Tellinidae includes the paper shells which is represented by a single genus. Tellinids are small benthic and deposit feeding bivalves (Ansell, 1961). The species of *Tellina* are common on all sandy shores. Tellinids are characterized by thin whitish or pinkish or yellowish white shells. The shell is usually more or less oval, sometimes elongated or pointed behind posteriorly (Gravely, 1941). The two valves are not alike, and the shell surface is smooth or irregularly sculptured. The ligament is external, the right valve usually has lateral teeth as well as cardinals. The pallial sinus is large, its lower margin often fused with lower part of pallial line. The gill plates are smooth. The foot is byssiferous. The long separate siphons may be several times longer than the length of the shell, and the inhalant siphon is nearly four times as long as exhalent (Gravely, 1942).

Tellina is a soft bottom, burrowing animal and shows many adaptations to suit its mode of life. The shell valves are modified, becoming more streamlined for burrowing by being flattened or elongated. Extensive mantle fusion not only contributes to the reduction of fouling but is probably equally important in maintaining the hydraulic pressure, within the mantle cavity necessary for burrowing. Mantle edges surrounding the apertures have become elongated to form tubular siphons of varying length. Since it burrows deeper it has long siphons. Foot bears a little flattened sole.

Tellinids are not suspension feeders as are most other bivalves. Thorson (1957) listed a number of soft bottom communities in which Tellina plays a significant role. *Tellina* burrows deeper than wedge shells. The most common and most widespread of all the deposit feeders is the *Tellina*. Recently shellfish fisheries have developed to such an extent that it plays a significant role in the Indian economy. The edible bivalve resources are tapped not only as food for the weaker sector of the coastal population but also used as a raw material in the preparation of lime which is used as a construction material. Bivalves are nutritious, delicious and abundant source of protein. Hornell (1922) reported the utilization of shells for manufacturing of lime. Clams like *M. meretrix,* M. casta, *Katelysia* opima, *Donax* spp. and the Cockles *Anadara* spp., and *Tellina* sp. are used as food by the poor people living in coastal areas (Hornell, 1949). Clams are highly esteemed for their taste and nourishment and so they play a dominant role in culinary of coastal people especially during monsoon when the fishes are not readily available (Durve, 1964). Despite their high delicacy, export potential and cheap availability, commercial culture of bivalves is still in its infancy in India. They can be cultured by simple and inexpensive method. There is vast scope for development of bivalve farming on a commercial scale in the backwaters and estuaries along the Indian coast. In Bombay and Madras, small sized oysters are collected and maintained in the field till they attain marketable size (Alagarswami, 1966). In India an area of 17 million hectares exists as backwaters and estuaries which could be fruitfully utilized for shellfish culture (Durve, 1964). The knowledge of biology and breeding habits is essential for culture production of any bivalves. Tellinids are at present consumed only by a limited numbers of people, and members of Tellinidae holds a good potential in pisciculture practices as it forms a good feed to fishes (Ansell and Trevallion, 1967). Although more than seven species of *Tellina* (Srihari, 1975) are available in Porto Novo waters, these species have never been the subject of any major investigation. Available information on the biology, breeding habits and nutritive values of the Tellinids *(Tellina* spp.) is inadequate.

1.2. DESCRIPTION OF THE STUDY AREA

The Vellar estuary, which has its origin in the Servarayan Hills in Salem District of Tamil Nadu, is located in latitude 11°29'N; longitude 67°60'E and it opens to the Bay of Bengal after traversing a distance of about 480 kms (Fig. 1). It has a variety of biotopes such as neritic, estuarine, backwater and mangrove regions. The Vellar estuary as an open type has connection with the adjoining sea throughout the year, is being subjected to semi-diurnal tides. The tidal range in the estuary is of the magnitude of 1.0 meter and its influence is noticed even beyond 16 kms from thc mouth. During northeast monsoon (Oct-Dec) this area experiences heavy downpour and Vellar drains more of freshwater into the Bay of Bengal. The average depth of the estuary is 2.5 meters, and the maximum is 5.0 meter. The estuary also receives discharge from a few irrigation channel namely long channel, dog channel, Buckingham channel and Railway bridge channel. Following Rochford's (1951) classification Ramamoorthy (1954) divided the Vellar estuary into four zones namely Marine zone, Tidal zone, Gradient zone and Freshwater zone. The nature of the bottom is sandy with black mud at surface depth. The Vellar estuary is always confluent with seawater and there is a complete closure of the bar, but the width, depth and position of the river mouth changes frequently due to the formation of sand bar across the mouth, and during the premonsoon and monsoon the estuary is directly connected with the sea (Pl. 1). Being a true estuary it is always subjected to long term fluctuations in salinity and other parameters resulting from seasonal variations in the of freshwater from rainfall. The shallow nature of the estuary facilitates rapid cycling of nutrients, thereby ensuring high productivity (Sivakumar, 1982). The flushing of the tide is relatively stronger so that the bar is always open.

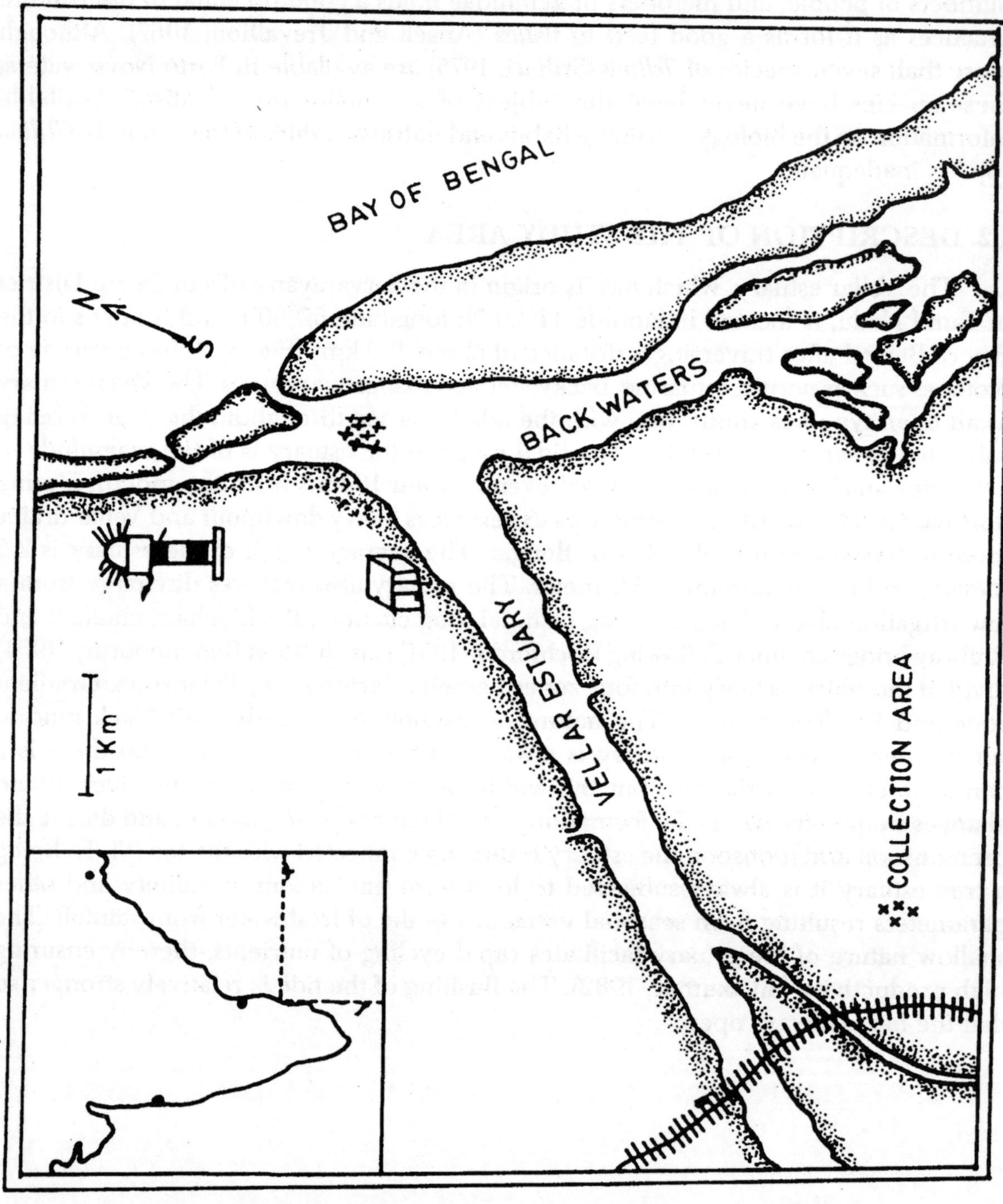

Fig. 1 : Map of Vellar estuary showing the collection area.

1.3. SYSTEMATIC POSITION

Phylum	-	Mollusca
Class	-	Bivalvia
Order	-	Eulamellibranchia
Sub order	-	Heterodonta
Series	-	Tellinacea
Family	-	Tellinidae
Genus	-	Tellina
Species	-	*Tellina ala* (Hanley, 1844)
		Tellina cuspis (Hanley, 1844)

1.4. SPECIES OF *Tellina* THAT OCCUR IN PORTO NOVO WATERS (Srihari, 1975).

Tellina rubra (Hanley)

T. *phillippinaram* (hanley)

T. *pristis* (Lam)

T. *timorensis* (Lam)

T. *nobilis* (Hanley)

T. ala (Hanley) and

T. *cuspis* (Hanley)

Table 1. Species of Tellina recorded from the coast of Tamil Nadu

Palk Bay Madras Beach Gravely, 1941) (Srihari, 1975)	Krusadai (Satyamurthi 1956)	Cochin Island (Satyamurthi 1956)	Portnovo waters Harbour area (Cheriyan, 1968)	
T. multangula	*T. ala*	*T. ala*	*T. rhodon*	*T. Rubra*
T. coarctata	*T. emarginata*	*T. angulata*	*T. pinguis*	*T. Phillippinarum*
T. angulata	*T. bruguieri*	*T. scalpellum*		*T. prishis*
T. emarginata		*T. bruguieri*		*T. timorensis*
T. nobilis		*T. coarctata*		*T. cuspis*
T. cuspis				*T. ala*
T. ala				*T. nobilis*
T. Pristis				
T. timorensis				
T. immaculata				
T. Papyracea				
T. pellucida				
T. rhodon				
T. texurata				
T. methoria				
T. rutila				
T. phillippinaram				
T. bruguieri				
T. pinguis				
T. lanceolata				
T. foliacea				
T. sinuata				

1.5. DESCRIPTION OF SPECIES : (Gravely, 1941)

T. ala: The shell is usually thin and moderately small, rather compressed and more or less transversely elongated. It is equivalved and pale white in colour. The front margin is more or less broadly rounded. The shells range from 2.2 mm to 33 mm in length and 1.1 to 16.5 mm in breadth. The shell shape is highly adaptive for burrowing movement with the anterior end undermost.

***T. cuspis*:**

Deep pink in colour, smaller than *T. ala*, upper margin of shell behind the umbo strongly arched and sloping steeply down. The shell range from 0.9 cm to 26 mm in length and 0.4 mm to 16.4 mm to width. It is very thin and fragile. It is a small Tellinacean.

T. phillippinaram

Upper hind margin arched and projecting shell smaller and thinner with regular microscopic concentric striae throughout the surface. Colour orange, pink shell only about one and half times as long as high.

T. pristis

Shell concentrically crested, hind margin toothed shell usually white, larger, radially furrowed behind.

T. timorensis

Upper hind margin not toothed, shell whitish and opaque. Surface of shell with more or less strongly developed, concentric growth lines, shell fairly high and rounded and not appreciably elongated transversely.

T. rubra

Shell pinkish, brittle in nature, slightly keeled from umbo to hind angle which is less rounded.

T. nobilis

The shell is white and elongated than *T. ala*. The shell shape is highly adaptive for burrowing movements.

1.6. DISTRIBUTION OF TELLINIDS

Gravely (1941) has recorded 22 species of Tellina from the shells occurring in Madras beech. Satyamurthi (1956) has recorded three species of Tellina from Palk Bay and five from Kurusadai islands Cheriyan (1968) has reported two species of Tellina, namely *T. rhodon* (Hanley) and *T. pinguis* from Cochin Harbour area. Seven species of Tellina have been reported from Porto Novo waters and two species *Tellina ala* and *Tellina cuspis* occur in the estuary.

1.7. REVIEW OF PREVIOUS WORK

Some of the important earlier studies on growth, reproduction, and biology of bivalves are those by Orton (1926) on *Cardium edule,* by Stephen (1928, 29) on *Tellina tenuis,* by Newcombe (1936) on *Mya arenaria,* by Coe (1947) on *Tivela stultorum* by Holme (1961) on *Tellina tenuis,* by Allen (1958) on *Nucula turgida* and *N. moori,* by (Ross Pohlo, 1957) on *Donax gouldi* by Segerstrab (1960) on *Macoma balthica,* by Penzies (1969) on *Tellina martinii,* by Hughes (1970, 1971) on *Scorbicularia plana,* by Guy Bachelet (1980) on *Macama balthica* and by Morton (1974) on *Musculista senhausia.* The works of Gravely (1941) and of Satyamurthi (1956) are helpful as general guides for identification of littoral molluscs of Madras coast. Recent studies are by Mclachlon and Hanekom (1979) on *Donax sordidus,* by Brook (1980) on *Cardium* edule, by Salzwedel (1980) on *Tellina fabula* and by Goshima (1982) on *Mya arenaria.*

Many informations are available on the distribution, ecology, reproduction, growth and biochemical aspects of the bivalves of Porto Novo waters. Studies on the anatomy and physiology of *Sanguinolaria diphos* were carried out by Venugopalan (1957). Krishnamurthy (1964) made a study on the distribution of iron in the tissues of some bivalve molluscs occurring in Vellar estuary. The biology of the venerid clam *Anadara rhombea* (George John, 1980) *Meretrix casta* (Balasubramaniyan, 1984), *M. meretrix* (Jayabal, 1984). *Katelysia opima* (Kalyanasundaram, 1986), *Sunetta scripta* (Diravidamani, 1986) and *Crassostrea madrasensis* (Marimuthu, 1986) were carried out recently. Among these investigations majority of works were carried out in the family veneridae.

Many notable works are available on the biology and functional morphology of the members of the family Tellinidae of foreign waters. Stephen (1928) described the biology of *Tellina tenuis.* He also studied the rate of growth of *Tellina tenuis.* Stephen (1928, '29) described the growth rate, mortality and breeding season of the intertidal and sublethal *Tellina tenuis* in Scotland and discussed the effect of tidal level of growth, Yonge (1949) described aspects of the anatomy and function and adaptations of various species Tellinacea, the deposit feeding eulamellibranchs. Thorson (1957) listed a number of soft bottom communities in which Tellina plays a significant role. Holme (1961) described the burrowing of several members of the Tellinidae and their habit of lying on the side in the sediment. Notes on the mode of life of the Tellinidae was studied by Holme (1961). McNulty *et al* (1962) studied relative abundance of *Tellina. Tellina* and its related genera occurring in the Western Atlantic were studied by Boss (1966). Functions of the shell ornamentation of *Tellina fabula* described by Wilson (1979) Guy Bachelet (1980) observed growth and Recruitment of the Tellinid *Macoma balthica.* Ansell and Trevallion (1967) studied the seasonal changes in biochemical composition of *Tellina tenuis.* Ross Pohlo (1967) studied the evolution of the Tellinacea. Wilson (1978) studied upper temperature tolerances of Tellina; A note on the synonymy of *Tellina subtrigona* was reported by Boss (1966). Population dispersion in *Tellina tenuis* Dacosta was reported by Holme (1961). Although seven species of Tellina have been reported from Porto Novo waters (Srihari, 1975) but these species have never been the subject of any major investigation and hence the present study was undertaken.

1.8. SCOPE OF THE WORK

The first chapter deals with general introduction, the aim of the study, biological significance of bivalves, description of study area, characteristic features of the family Tellinidae, systematic position, distribution of Tellinids along the coast of Tamil Nadu, species of Tellina which occur in Porto Novo waters, a review of previous work, adaptations of the burrowing in Tellinids and description of Tellina species.

The second chapter deals with ecology. There is very little information available regarding the salinity and temperature tolerances of benthic bivalves. Salinity and temperature tolerances experiments were carried out in both species by subjecting these two species to different levels of these two parameters. Since information available on the ecology of Tellina from Indian waters is scarce, the present study was carried out.

The third chapter deals with the morphology and anatomy of *Tellina ala* and *Te11ina cuspis,* knowledge of the animal anatomy is essential in studying its interrelationship with allied species and its environment. While Graham (1949), Owen (1955), Yonge (1949), Dinamani (1957) and Ansell (1961) studied the functional morphology of 12 species of Veneridae from British waters but there have been no such studies on Tellinids from Porto Novo waters. The fourth chapter presents the breeding biology of Tellinids which is essential for culture aspects.

Seasonal variations on biochemical constituents of Tellinids are presented in the fifth chapter. Tellinids are key members of the fauna of sandy shores and are possibly a major item in the diet of young fish in commercial fish rearing ventures (Ansell and Trevallion, 1967). Storage of biochemical nutrients becomes essential, so that they can be utilized, during the active reproductive period whether environmental conditions are favourable or unfavourbale to reproduction (Giese *et a1.*, 1967). During the annual reproductive cycle, intertidal invertebrates are exposed to different environmental conditions which may or may not favour reproduction. Seasonal variations on biochemical constituents on various bivalves were extensively studied (George John (1980) studied seasonal variations in the biochemical constituents of *Anadara rhombea;*

Jayabal (1984) on *Meretrix meretrix* and Balasubramaniyan (1984) on *Meretrix casta*). Investigation on the biochemical composition of various bivalves on a seasonal basis would be more informative. In the present investigation, an attempt has been made to study seasonal variations in protein, carbohydrate and lipid in the paper shells *Tellina ala* and *Tellina cuspis*.

The sixth chapter deals with effect of pesticides endosulfan (organochlorine) and dimethoate (organo-phosphorus) on the physiological activities like rate of filtration and respiration of Tellinids. Many investigations were carried out on the effects of heavy metals and pesticides on the physiology of various organisms but information on bivalves is scarce.

The seventh chapter deals with histopathological effect of the pesticide endosulfan on *Tellina ala*. Little information is available, on histopathological effect of pesticides. Study of histopathology of a pollutant is an important basic effort leading to our understanding of true impact of pollutants on the ecosystem.

2

ECOLOGY

2.1. INTRODUCTION

The distribution, growth, breeding and physiology of a species inhabiting an area are influenced by various ecological factors like temperature, pH, dissolved oxygen, salinity, availability of nutrients and substrate in its environment. Knowledge of the environment is, therefore, essential to understand the factors that control the distribution pattern of organisms (Shanmugam, 1987). Environmental factors affecting benthic communities have been studied by several authors and the more important works dealing with bivalves are those published by Moore and Lopez (1970), Hughes (1970); Rhoads (1970, '74), Young and Rhoads (1970), Bloom *et al* (1972). Kurian *et al* (1975), Richard Ivell (1979); McLachlan and Haneckom (1979); George John (1980); Kalyanasundaram (1982) Jayabal (1984) and Balasubramanian (1984) Hughes (1971) studied reproduction of *Scrobicularia plana;* Rhoads, (1970) reported mass properties, stability and ecology of marine muds related to burrowing activity; animal sediment relation in Cape Cod Bay was reported by Young and Rhoads (1971), Kurian *et al* (1975) studied ecology of bottom fauna of the Vembanad lake; McLachlan and Hanekom (1979) observed the aspects of the biology; ecology and seasonal fluctuations in biochemical composition of *Donax serra* in the East cape.

Factors controlling marine and estuarine bivalves have been reported by Wildish (1977). He reported that factors like salinity, oxygen, temperature and food availability influence the distribution pattern collectively. George John (1980), Kalyanasundaram (1982), Jayabal (1984) and Balasubramanian (1984) studied the ecology and biology of the venerid clams *Meretrix meretrix, M. casta*, *Katelysia opima* and *Anadara rhombea* from Porto Novo waters. They reported that sediment characteristics influence the distribution of bivalves in Indian waters.

There is very little information available regarding the salinity and temperature tolerance of benthic bivalves. Temperature tolerance experiments were carried out by Ansell *et al* (1980) on *Tellina fabula* and *T. tenuis*. He reported that *T. fabula* Gmelin, which is found in deeper water throughout its range, shows a lower thermal tolerance than *T. tenuis* da casta. Wilson (1978) has previously discussed the thermal tolerance of *Tellina fabula* and *t. tenuis* collected from Kames Bay, Millport in the Clyde Sea Area for LC 50. He also observed that *Tellina fabula* and *T. tenuis* showed well-marked differences in thermal tolerance.

Since information available on the ecology of Tellinids from Indian waters is scarce the present work was carried out to monitor the factors (salinity, temperature, pH, rainfall, dissolved oxygen, sediment analysis and population abundance) that control the distribution of *Tellina* sp. in Porto Novo waters.

These Tellinids experience fluctuation in salinity (0.6 to 37‰o) and temperature (28°C to 33°C) in their natural habitat. Experiments were, therefore, conducted to determine the tolerance limit of temperature and salinity to the benthic bivalves *T. ala* and *T. cuspis*.

2.2. MATERIAL AND METHODS

Present study on ecology relates to the factors like salinity, dissolved oxygen, temperature, food availability, rainfall and sediment analysis which appear to influence the biomass of Tellinids in Porto Novo waters. For the present investigator, the experiments were conducted in 3 replicates.

Hydrographical parameters

Data on monthly rainfall at Porto Novo during the present period of study was obtained from the meteorology section of our department run by the meteorological department of Government of India.

Surface water samples were collected using a clean plastic bucket. The temperature of surface water, and atmosphere were recorded using a mercury thermometer of ± 0.5 °C accuracy. For measuring sediment temperature the thermometer was inserted horizontally through the pit made for faunal sampling.

The pH (hydrogen ion concentration) of the water samples was measured by using a digital pH meter (ELICO model LI-10).

Salinity was estimated by the silver nitrate titration method as described by Strickland and Parsons (1968).

Dissolved oxygen was estimated following Winkler's method as described by Strickland and Parsons (1972). The texural class of the sediment was ascertained by plotting the values on a textural triangle devised by the United States Department of Agriculture (Anonymous, 1951). Portion of the sediment sample was powdered, passed through 85 u mesh sieve.

Animal collections were also made once a month to study the distribution and population density of the Tellinids in Porto Novo waters. Tellinids burrow rapidly in the moist sand and in some places, they are so abundant that they may be collected literally in handfuls by digging up the wet sand. Another method to capture bivalves is to sprinkle crab meat-dipped sea water. The bivalves emerge out of their burrows sensing the smell and then it is easy to pick them out. The local fisherman insert into the burrow a long and rigid midrib of coconut splinter and push it down carefully till it enters the mantle cavity of the animal. The insertion causes the animal to close the shell valves resulting the splinter firmly clasped between them. Animals are then dug out from the burrow from a depth of 2'6" to 3'0". Availability of other molluscs in the study area were also recorded.

Sampling with a grab or a dredge or sampling from a unit area with a help of a quadrate was not practical and so catch per unit effort was employed, as outlined by Smith (1971) for the population sampling in *Donax rugosus*. The same method was followed by George John (1980) on *Anadara rhombea* and Jayabal (1984) on *Meretrix meretrix* from Porto Novo coastal waters. Two or three persons were employed to hand pick the Tellinids for a period of three hours. The number and weight of the animals collected during each hour per person was calculated.

For salinity tolerance experiments, the animals were collected, brought to the laboratory and acclimatized to the laboratory condition for three days in their natural salinity of 28 ± 0.5 ppt before conducting the experiments. Different grades of salinities ranging from 0 to 33 ppt at intervals of 3 ppt were prepared by diluting seawater with appropriate volumes of dechlorinated tap water. For the lowest salinity tested (0‰) dechlorinated double distilled water was used. Ten animals of the same size group from each species were selected and placed for each salinity in glass troughs of 3 litre capacity filled with two litre of water of the desired salinity. The glass troughs were covered with a fine mesh net to prevent air contamination. The water was changed every 24hrs. The experiment was repeated thrice. The activity, survival and mortality of these Tellinids in different grades of salinity were recorded daily for 25 days, following the scale described by Arnold (1972) for gastropods (Table 4).

Aeration was provided to both experimental and control animals and the animals were fed with phytoplankton. Experiment on the temperature tolerance was conducted following the procedure of Fraenkel (1968) and Wolcott (1973). The animals were acclimatized and kept in glass troughs. During the experiment each animal was subjected to similar heat dosage. They were placed in water baths (habitat water) of desired temperature (20°C to 40°C at a difference of 2°C for 72 hrs). During the experiments the animals were fed with phytoplanktons.

2.3. RESULTS

The results showed that highest rainfall was recorded during the monsoon period and the lowest was recorded during the postmonsoon and summer seasons (Fig. 2). The monthly variations in rainfall is shown in Table 2. Highest rain fall (556.8mm) was recorded in the month of November and lowest rain fall (9.3 mm) was recorded in the month of June. The total rainfall recorded during the study period was 1792.1mm.

Fig. 2, showed the monthly variations in temperature of surface water, atmosphere and sediment during the study period. Variations were not much pronounced among atmospheric, surface water and sediment temperatures (0.5 to 1°C). However, the temperature was higher in the atmosphere than in surface water or sediment.

Generally the temperature was high (29°C to 33°C) during summer months and low (23°C to 25°C) during monsoon months (October to December). Monthly mean values of the temperature of atmospheric, surface water and sediment were presented in Fig. 2.

Fig. 2 showed variations in pH of water during the study period. The pH fluctuated slightly from 7.5 to 8. Maximum pH of 8.0 was recorded during the month of January 1988 and the minimum of 7.5 was recorded during September 1987.

A wide range in salinity was recorded in the study area. Salinity showed a decreasing trend in premonsoon and monsoon and an increasing trend towards summer. Lowest values of salinity (0.6 to 1.2‰) were observed during October to December and high values of salinity (32 to 37‰) during April to June were recorded. Fig. 3. shows the fluctuation in salinity for the period of one year.

Fig. 2 shows the seasonal variations in dissolved oxygen content of water which ranged from 4.5 to 7.0ml/l during the study period (June 1987 to May 1988). The highest oxygen content (7.0ml/l) was recorded during the monsoon (November 1987) and the lowest dissolved oxygen content (4.5ml/l) was observed during the summer (April 1988).

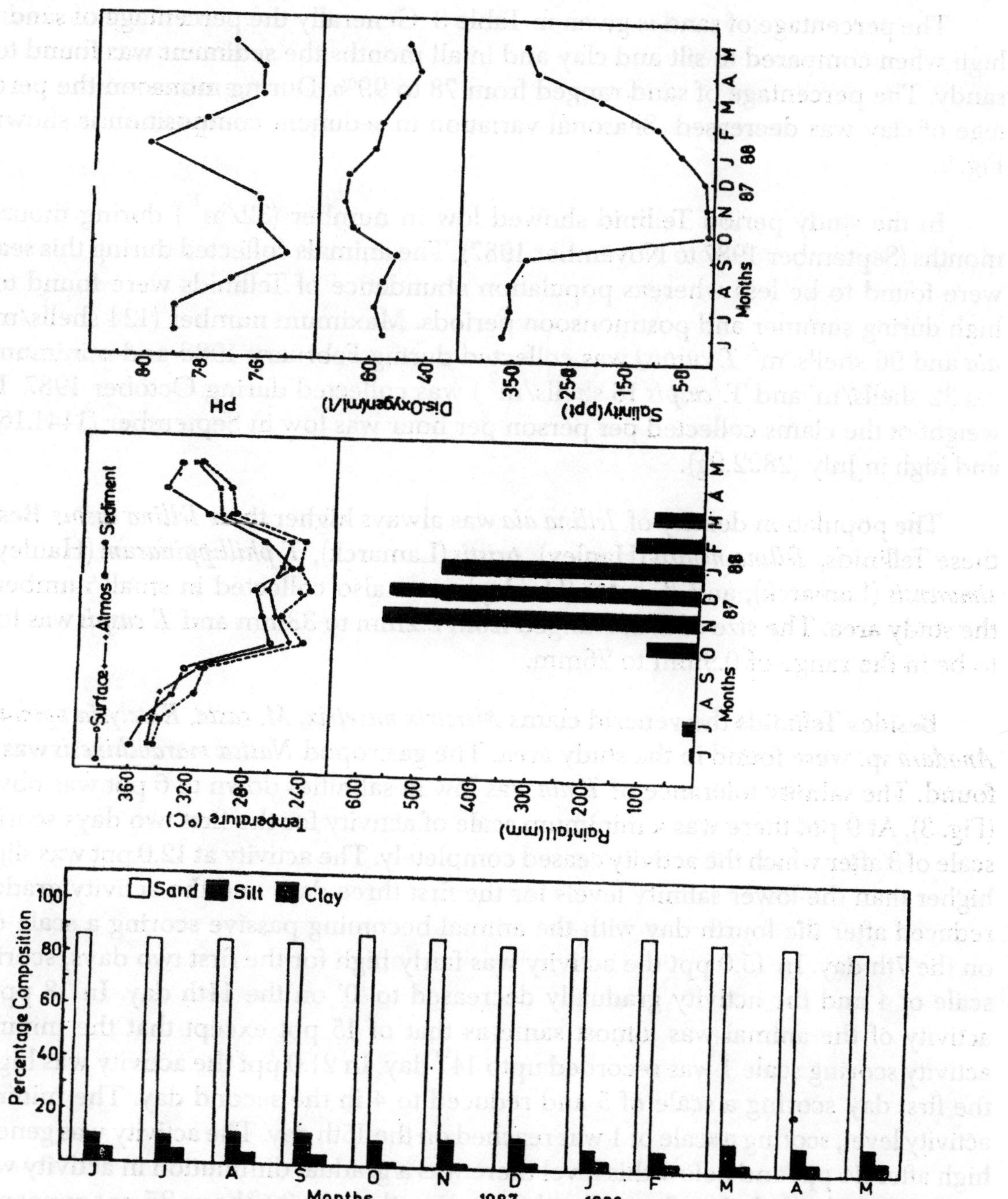

Fig. 2 : Diagram showing monthly variations in hydrographical parameters (Temperature, pH, dissolved oxygen, salinity, rainfall and percentage of sand composition.

The percentage of sand is given in Table 3. Generally the percentage of sand was high when compared to silt and clay and in all months the sediment was found to be sandy. The percentage of sand ranged from 78 to 99%. During monsoon the percentage of clay was decreased. Seasonal variation in sediment composition is shown in Fig. 2.

In the study period Tellinid showed low in number ($32/m^2$) during monsoon months (September 1987 to November 1987). The animals collected during this season were found to be less whereas population abundance of Tellinids were found to be high during summer and postmonsoon periods. Maximum number (124 shells/m^2 *T. ala* and 96 shells/m^2 *T. cuspis)* was collected during February 1988 and minimum (*T. ala* 32 shells/m^2 and T. *cuspis* 16 shells/m^2) was collected during October 1987. Total weight of the clams collected per person per hour was low in September (1441.16gm) and high in July (2822.9g).

The population density of *Tellina ala* was always higher than *Tellina cuspis*. Besides these Tellinids, *Tellina nobilis* (Hanley), *pristis* (Lamarck), *T. phillippinarum* (Hanley), *T. timorensis* (Lamarck), and *T. rubra* (Hanley) were also collected in small numbers in the study area. The size of *T. ala* ranged from 2.2mm to 33mm and *T. cuspis* was found to be in the range of 0.9mm to 26mm.

Besides Tellinids the venerid clams *Meretrix meretrix, M. casta, Katelysia opima* and *Anadara* sp. were found in the study area. The gastropod *Natica maraochinesis* was also found. The salinity tolerance of *T. ala* was low in salinities down to 6 ppt was obvious (Fig. 3). At 9 ppt there was a minimum scale of activity for the first two days scoring a scale of 3 after which the activity ceased completely. The activity at 12.0 ppt was slightly higher than the lower salinity levels for the first three days and the activity gradually reduced after the fourth day with the animal becoming passive scoring a scale of '0' on the 7th day. In 15.0 ppt the activity was fairly high for the first two days, scoring a scale of 4 and the activity gradually decreased to '0' on the 11th day. In 18 ppt the activity of the animal was almost same as that of 15 ppt except that the minimum activity scoring scale 1 was recorded upto 14th day. In 21.0 ppt the activity was high on the first day scoring a scale of 5 and reduced to 4 in the second day. The minimum activity level, scoring a scale of 1 was reached on the 15th day. The activity was generally high after 24 ppt and below this level there was a gradual diminution in activity which ceased completely below 9 ppt. For this species, the salinity above 25 ppt appeared to be the point at which it can undertake enough activity to acquire food and resist other adverse environmental conditions and 28-30 ppt was found to be the optimum salinity for *T. ala*.

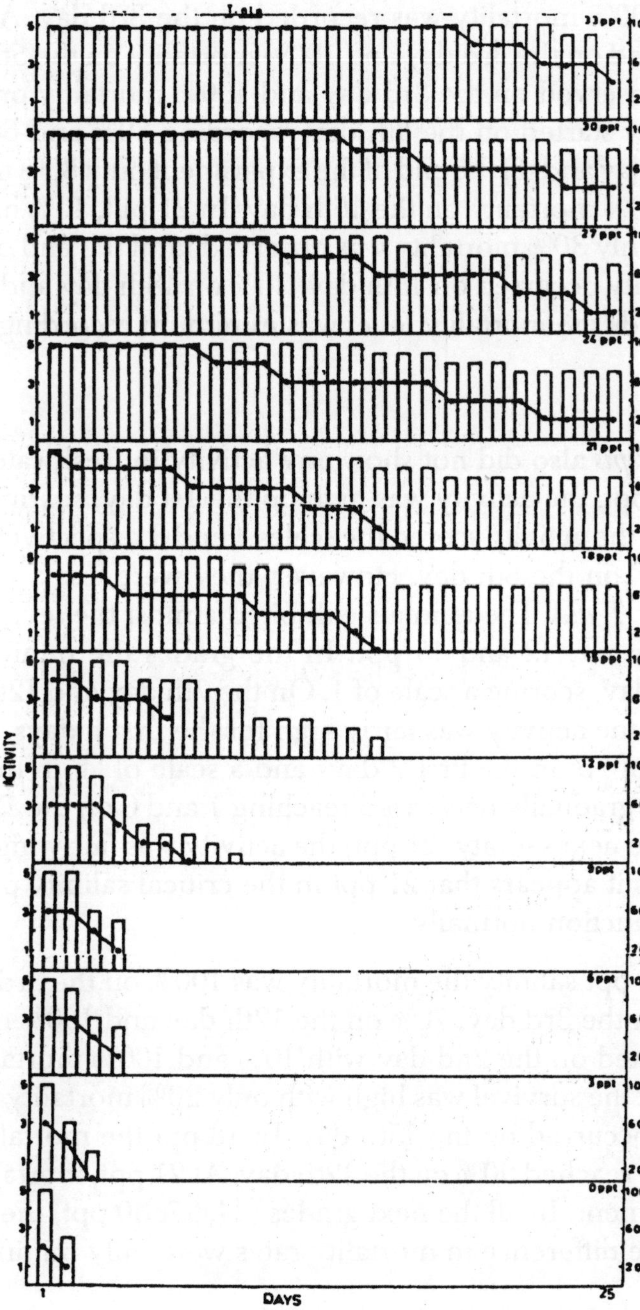

Fig. 3 : Diagram showing salinity tolerance of *T. ala.*

In fresh water 100% mortality was recorded on the 3rd day. At 6 ppt and 9 ppt 60% mortality occurred on the 3rd day and 100% mortality on the 5th day. At 12.0 ppt 40% mortality was observed on the 3rd day and 100% mortality on the 10th day. In 15.0 ppt, the mortality started on the 5th day recording 80% and 60% mortality was recorded on the 10th day and till the end of the experiment only 80% mortality occurred on the 7th day and 50% mortality on the 16th day. In 21 ppt, 10% mortality occurred on the 12th day and only 30% mortality was observed until the end of the experiment. In 24 ppt 10% mortality occurred on 15th day, 20% on 18th day and 30% on 21st day. From 24.0 ppt onwards the mortality level was minimum recording only 20%.

T. cuspis

Like *T. ala*, *T. cuspis* also did not show any activity in freshwater and upto 6 ppt salinity (Fig. 4). In 9 ppt, the activity was high on the 1st day, scoring a scale of 5. On the 2nd and 3rd days, the activity scored the scales 4 and 3 respectively and diminished gradually reaching '0' on the 8th day. However, they were less active at this salinity when compared with *T. ala*. The animals exhibited almost the level of activity in the two grades of salinity viz. 12 and 15 ppt. In the grades the minimum activity was observed on the 7th day, scoring a scale of 1. On the 7th day and 12th day the activity scored '0'. At 18 ppt, the activity was fairly high for the first 6 days. At this salinity *T. cuspis* scored a scale of ' 5' in the first 2 days and a scale of '4' in the next four days. After that the activity gradually decreased reaching 1 and 0 on the 12th and 15th days respectively. From the next salinity (21 ppt) the activity was fairly high until the end of the experiment. Thus it appears that 21 ppt in the critical salinity point below which the animals cannot function normally.

Between 0 and 6 ppt salinity the mortality was 100% on the 3rd day. In 9 ppt the mortality was 30% on the 3rd day, 70% on the 12th day and 100% on 13th day. In 12 ppt the mortality started on the 2nd day with 10% and 100% mortality was occurred on 15th day. In 15 ppt the survival was high with only 20% mortality until the 10th day and 100% mortality occurred on the 15th day. In 18 ppt the mortality started on the 7th day with 10% and reached 90% on the 12th day. At 21 ppt, mortality was 40% until the end of the experiment. In all the next grades (24, 27, 30 ppt) the survival rate was high (80-90%) and the difference in mortality rates were only minimal.

Temperature tolerance

When subjected to the temperature range of 20°C to 40°C both the species of *Tellinids* responded in a similar manner. However, *Tellina ala* exhibited slightly more tolerance to temperature than *T. cuspis* when the temperature was increased from

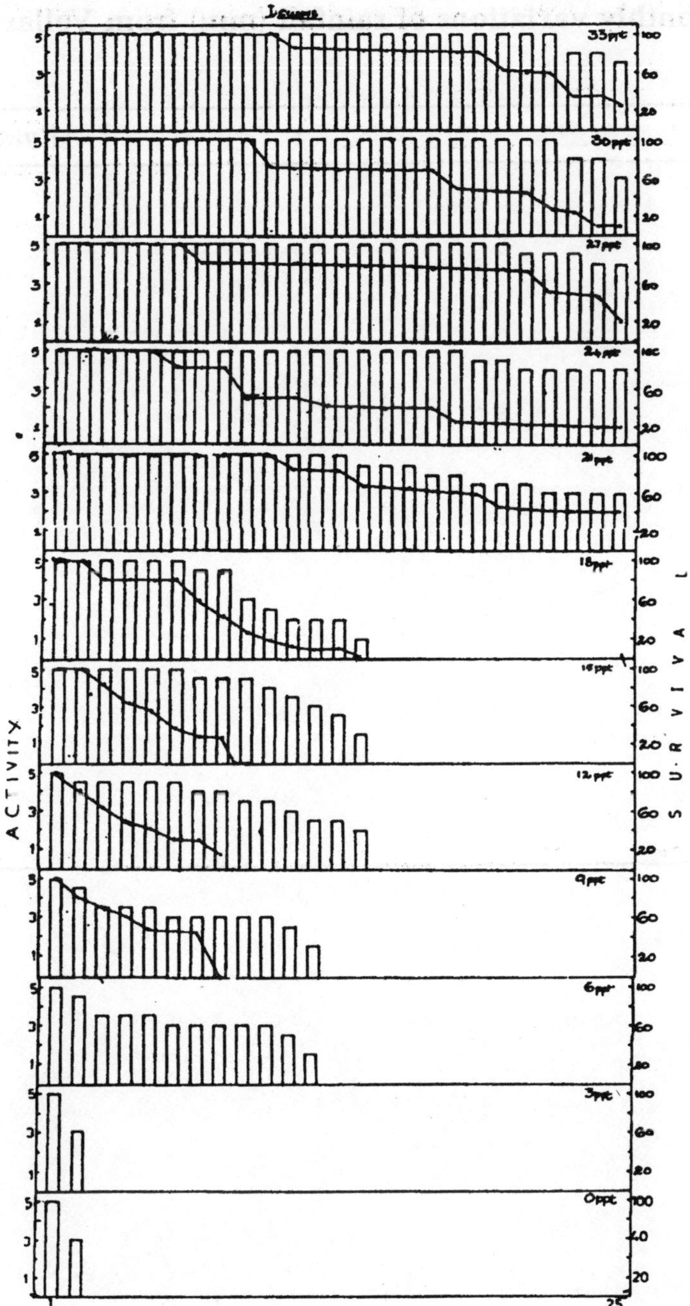

Fig. 4 : Diagram showing salinity tolerance of *T. cuspis.*

Table 2. Monthly variations of rainfall (mm) from Vellar estuary.

Months	*Rainfall (mm)*
June 1987	9.3
July	20.3
August	–
September	–
October	90.1
November	556.8
December	449.3
January	456.2
February	120.1
March	90.0
April	–
May	–

Table 3 : Monthly percentage of sand, silt and clay of sediments from Vellar estuary.

Months	*Sand* %	*Silt* %	*Clay* %
June 1987	86.0	9.5	4.5
July	85.3	9.7	5.0
August	84.6	11.5	3.9
September	84.4	11.9	3.7
October	87.9	10.0	3.5
November	87.0	11.2	2.9
December	84.6	11.5	2.5
January	89.0	7.5	2.0
February	88.9	8.2	1.8
March	87.0	10.5	2.9
April	85.9	9.9	4.0
May	84.5	9.8	4.7

Table 4. Scale of activity level in *T. ala* and *T. cuspis* (Arnold, 1972)

Details	*Score activity*	*Designation of*
Animals fully withdrawn into the shell	0	Inactive
Animals protruding foot and siphon but not moving; attached to the bottom	1	Low
Animals protruding foot, siphon and moving very slowly	2	Low
Besides protruding foot and siphon animals are attached to bottom of glass trough	3	Moderate
Animals attached to the bottom but not moving	4	High
Animals actively moving here and there below or above water level	5	High

20°C to 22°C. In this temperature *T. ala* recorded no mortality but there was 30% mortality in *Tellina cuspis*. Another 2°C raise in temperature resulted in 50% mortality in *Tellina cuspis* and 30% mortality in *Tellina ala,* 100% mortality was observed in *Tellina ala* at 40°C and at 38°C in *Tellina cuspis.* Thus the optimum temperature for both these species seems to be between 28°C and 32°C and above these temperature range both species showed mortality. However, between these temperature ranges both species were active enough to carry out their life sustaining activities like feeding, burrowing, etc.

2.4. DISCUSSION

The different hydrographic parameters observed in the present study showed wide fluctuations which could be attributed to the inflow of freshwaters during the monsoon. The mouth of the Vellar estuary, where the Tellinids were abundant is most dynamic due to the influence of neritic sea waters and freshwater inflow during the monsoon. The fresh water inflow which the mouth experiences in monsoon greatly reduced the salinity, temperature and changed the sediment composition which in turn greatly influenced the faunal distribution. Generally the sediment is composed primarily of sand at all times. During monsoon the percentage of clay was very low. However, small seasonal variations in clay content could be a result of increased turbulance during the monsoon which would reduce sedimentation rate and probably remove some from the sediment. During summer neritic water enters the mouth of the Vellar estuary effectively increasing salinity, temperature, nutrient levels and influencing the faunal distribution.

In the present study temperature was higher in summer and postmonsoon season and lower in monsoon and premonsoon this suggests seasonal changes in solar radiation. The changes in water temperature may be affected more likely by increased fresh water runoff of lower temperature than that of the neritic waters. Similar observations are also reported by Jacob and Rangarajan (1959); Dyer and Ramamoorthy (1969); Vijayalakshmi (1973); Rajendran (1984) and Ramadhas (1977) in the Vellar estuary and McLusky *et al* (1975) from a Central West Indian beach and Achuthankutty *et al* (1978) from the west coast of India. A fairly close relationship between the atmospheric and water temperature was observed in the present study. This phenomenon was reported by Rangarajan (1958) and Vijayalakshmi and Venugopalan (1973) in the Vellar estuary. In general, temperature is an important factor in the distribution of animals (Kinne, 1967).

In the present study, hydrogen ion concentration showed a narrow range (7.5 to

8). (No significant correlation could be established between the pH and distribution of animals.)

During the present study, maximum salinity was recorded in summer months and minimum in monsoon months. This may be due to the increase in the influence of seawater and low salinity in monsoon may be due to heavy rains and increased fresh water flow. A similar condition was earlier reported by Dwivedi *et al.* (1973) in Mondovi estuary, Ansell *et. al.* (1980), Achuthankutty *et al* (1978) and Nair (1978) at different open sandy beaches along the East coast of India by Olivia (1981) and Sivakumar (1982) in Vellar estuary.

Salinity seems to play a major role in controlling the distribution of benthic animals (Underwood, 1979). Influence of salinity on the distribution of stenohaline and euryhaline bivalves was detailed by Dodgeson (1928), Jutting (1943), Arnold (1972), McErlean (1964), Lassig (1965), Kinne (1967) and George John (1980), Kurian *et al* (1975) and Nair (1978) also commented on the possible role of salinity on the distribution of brackish water bivalves. Mass mortality of *Nodiolus metcalfi* was recorded during monsoon due to the sudden decrease in salinity in Cuddalore water by Marimuthu (1984).

The salinity of the coastal water is influenced by freshwater runoff from the land and by incursion of neritic waters. Decrease in salinity was usually observed during monsoon and in other seasons. However, the salinity is almost constant or at an optimal level due to the influence of neritic waters in Vellar estuary during summer (Kanagasabai, 1985). High salinity coincided with high temperature values, during summer months. During summer, the estuary was dominated by neritic waters and during monsoon by freshwater.

In the present study, it was observed that dissolved oxygen content was higher in premonsoon months and lower during summer which is in agreement with results recorded by Dye (1977) in an estuary from South Africa. Heavy rainfall during October to December caused heavy flow of freshwater which might be the reason for higher values in dissolved oxygen content recorded during monsoon. Similar observations were reported by Krishnamurthy (1964), Vijayalakshmi (1973), Jagathesan (1974) and Rajendran (1984) for Vellar estuary. Seasonal variations on dissolved oxygen content of the Vellar estuary have earlier been reported by Olivia (1981) and Sivakumar (1982).

Redding and Cory (1975) reported that in environments with nearly identical temperature and salinity regimes, sediment characteristics might reasonably be considered as an important factors influencing the distribution of benthic fauna. Sanders (1958) considered that, the sediment characteristics greatly influenced the distribution of various benthic organisms in Buzzards Bay, U.S.A. Ansell and Trevallion (1967), Wolff (1973); Hibbert (1977) and Broch (1979) also considered that the sediments and substrate influenced the distribution of bivalves.

It was observed that the abundance of Tellinids decreased during monsoon season and increased during summer season. This may be due to heavy rainfall during monsoon and the consequent sudden lowering of salinity. Sudden mortality of mussels were recorded during monsoon (Marimuthu, 1984) due to lowering of salinity. Seasonal changes including rainfall may influence the density of population of intertidal areas (Odum, 1963). Decrease in Tellinids population was observed during monsoon as a result of sudden environmental changes. In the present study the population abundance of *T. ala* was dominant than *T. cuspis*. The dense population of *T. ala* and *T. cuspis* was observed during April 1988 to May 1988 and minimum number was collected during premonsoon month of August and during monsoon months of September to November 1987.

The mouth of the Vellar estuary experiences fluctuations in salinity, temperature, strong wave action, water current, substrate composition and faunal distribution. In the present study, the rainfall during monsoon months brought in heavy freshwater inflow and as a result there was a steep decline in salinity and temperature. The salinity decreased during the monsoon months which in turn influenced the temperature. The substrate analysis also showed a change due to the large amount of silt brought in by the freshwater runoff. Thus the monsoon and the successive rainfall has a greater influence upon the Tellinids found in the mouth region of the Vellar estuary.

Salinity tolerance

Underwood (1979) attributed rainfall as the main reason for the reduced salinities that the intertidal animals experience during monsoon months in tidal pools and coastal waters. During monsoon due to heavy freshwater runoff, the distribution of intertidal species are considerably reduced (Tapping and Fuller, 1942). Salinity variations do seem to affect the distribution of molluscs as noted by Newell (1964) and Frenchel (1968). The major role played by salinity as a factor controlling the distribution of intertidal animals has been brought to light by Underwood (1979). In tropical waters, salinity coupled with temperature is known to play a key role in the distribution of nearshore animals where the fluctuations in the salinity are well pronounced (Ajmal Khan and Natarajan, 1981). Intertidal snails adopt themselves to changes in salinity caused by evaporation and rainfall during summer and monsoon respectively (Hyman, 1967).

Informations available on the salinity tolerance of bivalves particularly Tellinids are almost absent. In the present experimental study *T. ala* showed normal activity in the salinity range of 23-26 ppt and in *T. cuspis* the range was 27-30 ppt. Similar observations were made by Arnold (1972) and Thivakaran (1988) in Littorinids and found that salinities below 25 ppt were not suitable to this species. In the present study 100% mortality of Tellinids were observed below 24 ppt. Many investigations were carried out on salinity and temperature tolerance in gastropods especially in Littorinids. These studies showed that there was a rapid decline in activity levels of *Littorina saxatilis* and *L. littores* at salinities lower than 20 ppt and the activity was too low at a salinity of 12 ppt. In the present study Tellinids showed no activity upto 6 ppt and below 24 ppt the activity is very low scoring a scale of 1 and 2. At 28 ppt the activity was maximum scoring a scale of 4 and 5. In the present study area *T. ala* has a distributional limit which is higher than *T. cuspis* and they were found to have a tolerance to salinity fluctuations.

Temperature tolerance

Temperature and salinity are the factors that are known to influence the distribution of animals in general (Kinne, 1967). Temperature tolerance of Tellinids and some European molluscs were studied by Ansell *et al* (1980). They observed that the effect of reduced salinity is to lower the thermal tolerance especially for *Donax vittatus.* This is in agreement with the general principle that fluctuations in combinations of environmental stress will combine to lower tolerance limits {Kinne, 1972). A similar observation was made in the present study and at low salinity, the thermal tolerance was reduced in *T. ala* and *T. cuspis.* Ansell *et al* (1980) observed that thermal tolerance of *Donax* sp. are in accordance with their distributions. *T. ala* by virtue of its high tolerance to temperature and salinity, generally occupied more readily in numbers than T. *cuspis.*

3

MORPHOLOGY AND ANATOMY

3.1. INTRODUCTION

Knowledge of the animal anatomy is essential in studying its interrelationship with allied species and its environment. Anatomical studies on bivalves were mostly confined to body structures like ligament, gill, stomach, etc. (Atkins, 1937; Graham, 1949; Owen, 1955; Yonge, 1949; Dinamani, 1957, '67; Ansell, 1961). Ansell (1961) described the functional morphology of 12 sp. of Veneridae from British waters and Sellmer (1967) dealt with the anatomy of the gem clam, *Gemma gemma* from New Jersey, U.S.A.

Notable works dealing with the digestive system of bivalves are those of Yonge (1923, '26b, '49) on *Mya ostrea* and *Tellina*. Yonge (1926) also described the structure and physiology of the organs of feeding and digestion in *Ostrea edulis,* Venugopalan (1955) on *Sanguinolaria diphos,* Dinamani (1957) on *Villorita,* Galstoff (1964) on *Crassostrea,* Judd (1979) on *Macoma* and Morton (1969) on many bivalve species.

A comparative study by Graham (1949) on the structure of the stomach in 16 species of bivalves has formed the basis for later studies (Alien, 1954, '58; Dinamani, 1957; Owen, 1955; Purchon, 1955, 57, '58, '60, '77). In India, Dinamani (1967) made a comparative study on stomach of 36 species of bivalves. Purchon (1958) extended this to cover the stomachs of 60 genera of bivalves including Tellinids. The structure and arrangement of the ducts of the digestive diverticula of some Tellinids have been studied by Owen (1955, '66, '59). The morphology of the bivalve crystalline style, was examined by Nelson (1918, '25), Edmondson (1920), Yonge (1923, '26b, '49) and Shahul Hameed (1984).

Most of these studies have been carried out on the digestive system of various bivalves. However, no attempt has been made to study the general anatomy of Tellinids available in Indian waters. In view of this lacunae existing in our knowledge on Indian Tellinids, this present attempt has been made to study anatomy of two Tellinids namely *T. ala* and *T. cuspis.*

3.2. MATERIAL AND METHODS

Tellina ala and *T. cuspis* were collected from the mouth of the Vellar estuary and kept in circular troughs filled with filtered estuarine water. The animals were obtained by breaking open the shell. After removing the flesh part, the shells were cleaned using a brush with washing powder to carry out further studies. The morphometrical characters of the shell including shell length, and width were measured by using a Vernier calliper to a nearest of 0.1 mm level. For anatomical work, freshly killed as well as formalin/glycerol preserved animals were used. The anatomy was studied by dissections and by serial sections of whole animals and of various body organs. Camera lucida drawings, were made with the kelp of a stereozoom microscope. For serial sections, specimens ranging in length from 8 to 12 mm were used. The fixatives employed were Bouin's fluid and Zenker's fluid. Fixed tissues were embedded in paraffin wax, sectioned at 6-8 u thickness and then stained with Delafield's Haematoxylin with Eosin as counterstain. At each time for the study of morphology and anatomy 50 animals were sacrificed.

3.3. RESULTS AND DISCUSSION

3.3.1. SHELL

Tellina ala

The shell is usually thin and moderately small, rather compressed and more or less transversely elongated. It is equivalved and pale white in colour (Pl. 2). The outer surface of the shell is smooth except for the concentric lines of growth. The ligament is external. The lower end of the posterior margin of the shell is almost pointed angular. The front margin is more or less broadly rounded (Fig. 5). The left valve bears a single cardinal tooth while two occur on the right, the right valve usually having elongated lateral teeth in addition. The pallial sinus is very large, deep and extensive and its lower border is often found to coalesce with the portion of the pallial line immediately below it. The shells range from 2.2 mm to 33 mm, in length and from 1.1 to 16.5 mm in width. The shell shape is highly adapted for burrowing movements with the anterior end undermost, like in *Solen* and *Siliqua* (Pohlo, 1963).

Tellina cuspis

The shape and structure of the shell of *T. cuspis* is similar to *T. ala* except for colour and size (Pl. 3). The shell of *T. cuspis* is deep pink in colour, not radially furrowed behind. Upper margin of shell behind the umbo strongly arched and sloping steeply down (Fig. 6). The shell is without any radial depression about the middle. The size of *T. cuspis* is smaller than *T. ala.* The shell ranges from 0.9 mm to 26 mm in length and 0.4 mm to 16.4 mm in width. When compared to *T. ala*, the shell of *T. cuspis* is very thin and fragile *T. cuspis* is a smaller tellinacean.

Fig. 5 : Outer and inner view of shell valve of *T. ala.*

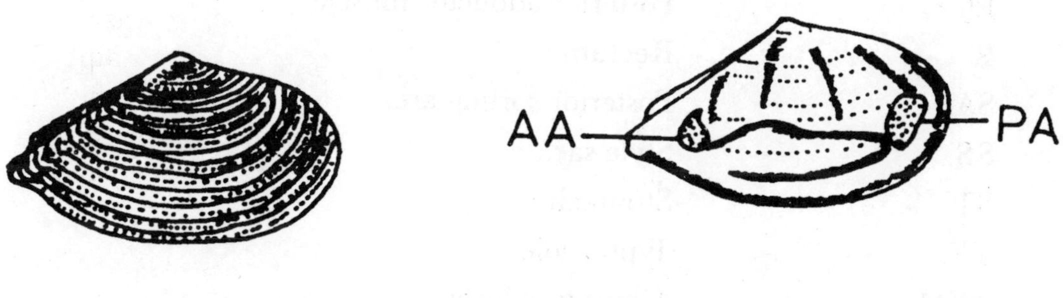

Fig. 6 : Outer and innver view of shell valve of *T. cuspis*

LIST OF ABBREVIATIONS USED IN THE FIGURES (8-14)

AA	-	Anterior adductor muscle
C	-	Crystalline style
CM	-	Cruciform muscle
DH	-	Dorsal hood
DP	-	Dorsal pouch
ES	-	Exhalent siphon
F	-	Foot
DG	-	Digestive gland.
GS	-	Gastric shield
GSL	-	Gastric shield (left pouch)
GSD	-	Gastric shield
ID	-	Inner demibranch
IS	-	Inhalent siphon
LP	-	Labial palp
M	-	Mouth
MA	-	Mantle
OD	-	Outer demibranch
OL	-	Outer lobe of mantle edge
P	-	Periostracum region
PA	-	Posterior adductor muscle
R	-	Rectum
SAP	-	Posterior sorting area
SS	-	Style sac.
ST	-	Stomach
TY	-	Typhlosole
TYM	-	Minor typhlosole
TYT	-	Major typhlosole
MG	-	Mid gut
ATY	-	Anterior typhlosole

3.3.2. SHELL MUSCLES OF TELLINIDS (adductor muscles)

The muscle impressions on the inner side of the shell include (i) the anterior adductor impression which is irregularly triangular and lies below the anterior extremity of the hinge, (ii) the impressions of the three elevator muscles lying *a* little above the adductor impression and (iii) the posterior adductor impression, which is the single largest muscle impression and is more or less 'D' shaped in outline (Fig. 6). The anterior and posterior impressions are connected by the pallial line which runs parallel to the edge of the shell (Fig. 6). The anterior and posterior retractor muscles which play an active role in digging process of the foot, arise above the adductor muscles and extend over the sides of the foot in the anterior and posterior region. There is no difference in the shell muscles of *T. ala* and *T. cuspis*. In the Tellinids the shell muscles in the pallial region of the mantle are employed for closing/opening of the shell. When the shell is closed, the edges fit closely together except at the posterior dorsal margin. Some water from the external medium can be retained within the pallial cavity and this may help the animal respire for a while, even when access to external aquatic medium is restricted.

3.3.3. MANTLE OF TELLINIDS (*T. ala* and *T. cuspis*)

The mantle lobes are thin, free from one another anteriorly but posteriorly they are fused. The pallial fusions separating the branchial and pedal orices are furnished with crosses of muscular bundles, the cruciform muscle, passing from the edge of one valve to the other. These form accessory adductor strands. The free edge of the mantle is much thickened and consists of three marginal lobes (Fig. 7 and 8). The outer one secretes all but the innermost layer of the shell. The middle marginal lobe bears a single row of branched tantacles and is bounded on its units outside by the periostracal groove, the area between this and the outer lobe being thus covered with periostracum. The inner lobe consists of a low musculature ridge which increases in size posteriorly. These inner marginal lobes which along unite posteriorly in the region of the cruciform muscle to form the mobile and separate siphons, the other two being continued a round the posterior margin of the shell bounding an embayment known as the siphonal space. The cruciform muscle which is situated at the base of the inhalent siphon represents an enlargement and specialization of the pallial musculature at the point of union of the inner mantle lobes (Fig. 7 and 8). The disposition of the organs in the mantle cavity as revealed after removal of the right shell valve and mantle lobe is shown in (Fig. 7 and 8). The extent of mantle fusion is similar in both species and typical of that found generally in group involving fusion of the inner folds posteriorly to form the siphons and progressive fusion dorsally culminating in the fusion of the

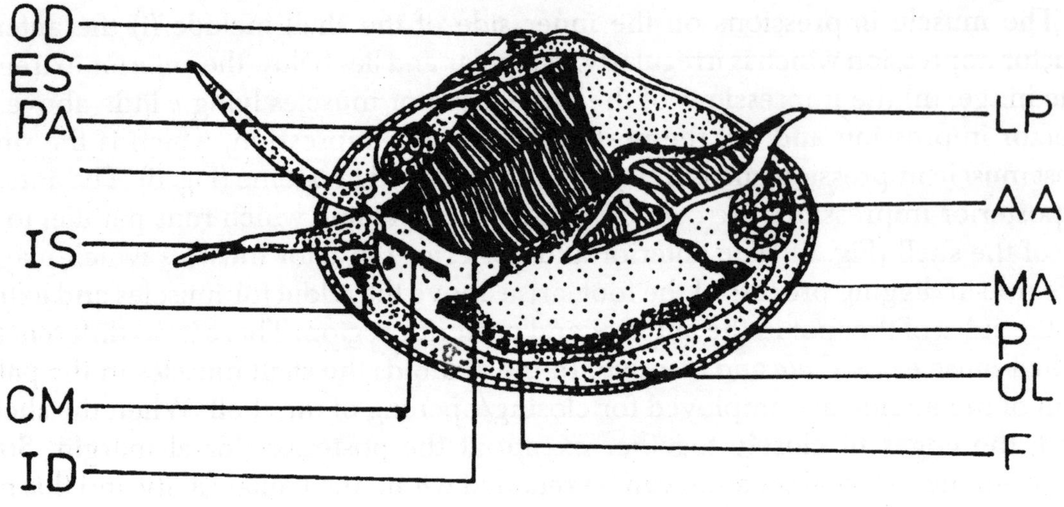

Fig. 7 : Diagram showing the internal organs of *T. ala*.

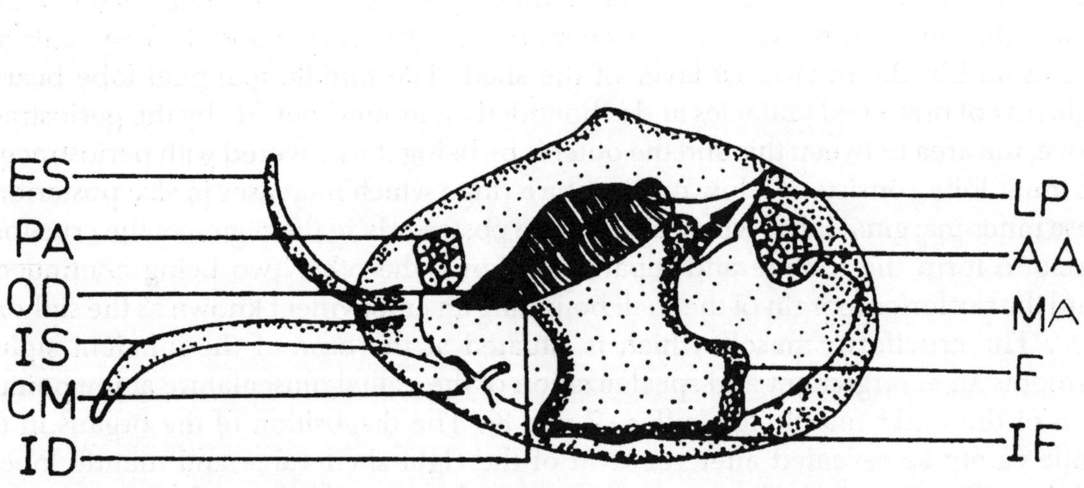

Fig. 8 : Diagram showing the internal organs of *T. ala*.

inner surfaces of the outer folds to produce a fusion layer extending to the ligament, especially posteriorly. The greatest differences between the two species lie in the structure of mantle edge and siphons and in the complexity of development of the tentacles, along the edges of the extensive pedal gape and at the tips of the siphons. The epithelium of the mantle lobe contains numerous glandular cells. The thickended edge of the mantle is full of connective tissue, muscle fibers and blood spaces.

3.3.4. MANTLE CAVITY IN TELLINIDS

The organs of the mantle cavity are shown in Fig. The demibranchs are large and elongated, while the labial palps are small. In the bivalvia the mantle cavity lies latero-ventrally between the left and right mantle lobes, and forms a spacious chamber, the principal function of which is to house the feeding and respiratory organs.

3.3.5. LABIAL PALPS IN TELLINIDS

The labial palps are paired triangular structure lying anteriorly at the oral end of the animal. There are two palps an inner and an outer palp on each side of the mouth (Fig. 7 and 8).

The demibranchs are quite large and the labial palps are small. Labial palps are found lying on the sides of the oral end of the animal. They are attached to the visceral mass by their broad base and their posterior free ends are narrow. The labial palps possess relatively large ridges on the opposed faces. The two labial palps on each side enclose a narrow groove between their basses. Further, the inner surface of each labial palp bears ridges which run across the width of the palp. The ridges especially towards the free margin of the palp, slope slightly in the anterior direction. The successive ridges are separated by furrows. The outer surface of the palp, which appears smooth, has low ridges with prominent cilia. These Tellinids resemble that of the deposit feeders by having upturned outer demibranch but it resembles suspension feeders in having large ctenidia and small labial palp.

3.3.6. THE CTENIDIUM

T. ala and *T. cuspis* resemble deposit feeders by having on upturned outer demibranch which is triangular in outline (Fig. 7 and 8).

On each side of the visceral mass, occur the outer and inner demibranch, each of which is composed of the usual filaments, inter filamentar junctions, inter lamellar junctions, ostia and water tubes as in other bivalves (Pl. 21 and 22). It lies entirely dorsal to the ctenidial axis and its ascending lamella is joined to the body wall overlying the pericardium and excretory organs. The inner demibranch is very narrow, and longer than the outer demibranch. It lies ventral to the gill axis and is thrown into much deeper folds than the other. The posterior extremities of the demibranchs are narrow and extend up to the region of the inhalent siphon. In general, ctenidia are relatively small and occupy less than a third of the extent of the mantle cavity (Yonge, 1949). In a cross section through the gill the arrangement of the principal filament and other filaments on each side is clearly observed (Pl. 7). Internally the filaments are supported by chitinous rods and the epithelium consists of a single layer of ciliated cells, the ciliation being retracted to the lateral, laterofrontal and frontal regions. Horizontal and vertical muscles are present as in all other eulamellibranchs. The ctenidium of *T. ala* and *T. cuspis* show much resemblance. Tellinids resemble suspension feeders by having relatively larger ctenidia than labial palp.

3.3.7. SIPHON OF THE TELLINIDS

As is characteristic of the Tellinacea (Yonge, 1949) the siphons of *T. ala* and *T. cuspis* are formed by fusion of the inner lobe of the mantle edge. They are long, separate and mobile (Fig. 7 and 8). The inhalant siphon is more extensive than exhalent siphon reaching a length of atleast four times that of the shells in *T. ala*. It is shorter in *T. cuspis*. The tip of the inhalent siphon bears small lobes. There are no true tentacles as are found in the inhalent siphons of suspension feeding lamellibranchs. These Tellinids resemble deposit feeders by having separate siphons which lack straining tentacles. In contrast to the great sensitivity of the siphons in suspension feeding bivalves, the tip of the siphons in *T. cuspis* and *T. ala* as in other Tellinacea, is very insensitive, seldom closing even when touched with a needle. Longitudinal and circular layer of muscles occur internally between the epithelium of the siphonal walls in *T. ala* and *T. cuspis* (Pl. 15). The entrance to the inhalent siphon has 6 lobes that are drawn out into finger like processes. There are no tentacles present on these lobes. The exhalent siphon of *T. cuspis* and *T. ala* has 6 lobes that are drawn out into finger like processes. There are no tentacle present on these lobes. The exhalent, siphon is devoid of lobes. The differences between the species lie in the structure of mantle edge and siphons, and in the complexity of development of lobes along the edges of the extensive pedal gape and at the tips of the siphons. Features of anatomy and behavior may be associated with the particular type of feeding.

3.3.8. FOOT IN TELLINIDS

The foot is relatively larger in *T. ala* than *T. cuspis*. It is somewhat elongated and axe-shaped in *T. ala* while in *T. cuspis*, the foot is slightly blunt (Fig. 7 and 8). The shape of the shell and foot are highly adapted for burrowing. Protractor pedis muscles are attached to the shell below the squarish insertion of the anterior adductor muscle and passing back superficially over the foot (Pl. 10). The fibres of the posterior retractors from the next layer, running forwards from oval insertions adjacent to those of the posterior adductor (Pl. 11). The innermost layer of all consists of strands of the anterior retractors which are inserted to each value by triangular insertions, the base uppermost, just posterior to those of the anterior adductor. The foot is used very actively for quick burrowing in Tellinids (Yonge, 1949). The foot is highly extendable and is provided with paired anterior and posterior pedal retractors. The epithelium of the foot consists of cubical cells and beneath the epithelium are seen several layers of foot muscles (Pl. 18). The connective tissue of the foot has numerous blood spaces and phagocytes (Pl. 18). The length of the foot when fully extended measure 6.2 ± 0.8 mm in *T. ala* and in *T. cuspis* 5.4. ± 0.6 mm. The width is 5.6 ± 5.6 ±0.5 mm in *T. ala* and 3.2 ± 0.5 mm in *T. cuspis*. The foot of *T. cuspis* appears to be smaller than that of *T. ala*.

Yonge (1949) indicated that the presence of separate inhalant and exhalent siphons, absence of true straining tentacles, the presence of mantle fold and upturned outer demibranch are features associated with deposit feeding bivalves.

Both *T. ala* and *T. cuspis* belong to same type of feeding (deposit feeders). There is not much difference in the anatomy except for morphological features like shell shape, size, colour and light variations in the structure of ctenidia and siphons.

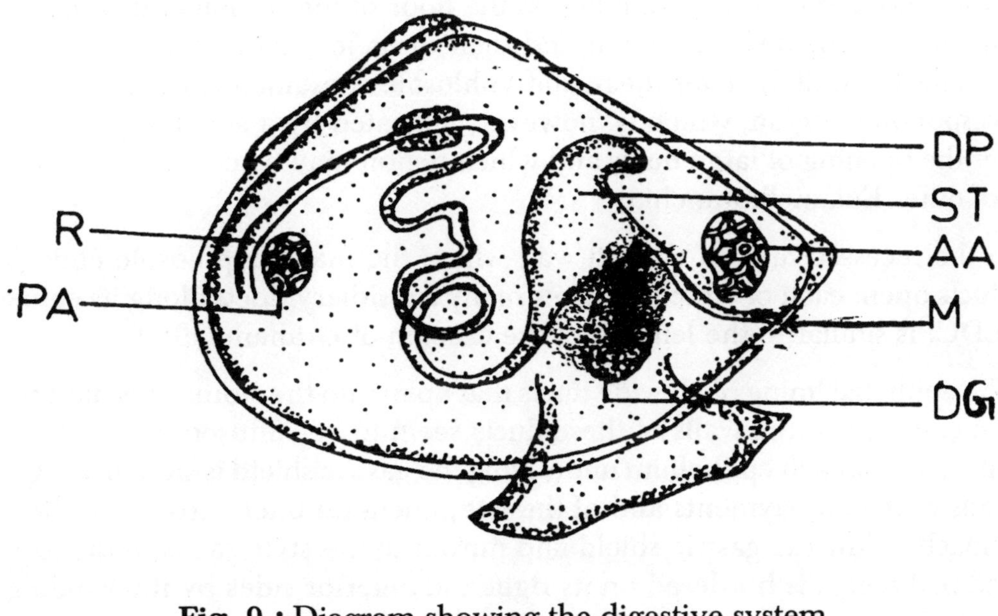

Fig. 9 : Diagram showing the digestive system.

3.3.9. DIGESTIVE SYSTEM IN TELLINIDS

Anatomy of digestive system of *T. ala* is presented in (Fig. 9; Pl. 4, 5, 6, 8, 9, 12, 13 and 14).

The mouth is a narrow horizontal slit situated behind and ventral to the anterior adductor. It is thus well protected from the entrance of harmful substances. It is flanked above by the thin upper lip, which is continuous with the outer labial palp on each side, and below by the lower lip which is continuous with the right and left inner labial palps. The most conspicuous features are the very capacious stomach while the oesophagus which is relatively short and open into the anterior part of the large stomach. The stomach consists of the usual two parts noted by Graham (1949) and Yonge (1949). The main variations in stomach structure are due to the interaction of the stomach walls with the duct system of the digestive diverticula. In *Tellina ala,* the stomach wall is thin and tough. Dorsally and posteriorly it is reinforced by a large and massive gastric shield (GS) (Fig. 10 and 11). Anteriorly at the left side is a well developed dorsal hood (DH) which curves deeply to the left and a pointed spur of the gastric shield (GSD) extends into it. A similar shield (GSL) is thrust open into left pouch, which is situated ventrally to the hood. The wide ducts open into left pouch with a duct giving off 3 or 4 subsidiary ducts. The minor typhlosole (TYM) ends as a wall prominence on the right wall immediately above the common opening of the midgut (MG) and the style sac (SS). The major typhlosole (TY) emerges from the midgut as a thin fold and curves to the right across the floor of the stomach. It runs anteriorly and dips into a shallow impression and then turns left and forms loop (TYT) in a recess on the left wall. In *T. ala* the major typhlosole does not form a conspicuous loop in corresponding region, which is, however, indicated by a small depression, which receives the opening of large ducts and which region may be considered as similar to those of other Eulamellibranchia.

In the recessed area of the left wall where the major typhlosole ends (TYT) 2 wide ducts open, each of which receives many subsidiary ducts along its course. This area (LDC) is similar to the left half of the caecum of Graham (1949).

The epithelial lining of all wide ducts that open into the stomach is dark brown or black in colour, and the walls of these ducts seem to be suffused with a dark brown pigment. The stomach epithelium underlying the gastric shield is similarly pigmented. The walls of the embayments and of the left pouch are finely grooved. The floor of the stomach below the gastric shield and infront of the style sac is in the form of an elevated pad which is bordered on its right and anterior sides by the winding major typhlosole. To the right of the major typhlosole and occupying the region of the right

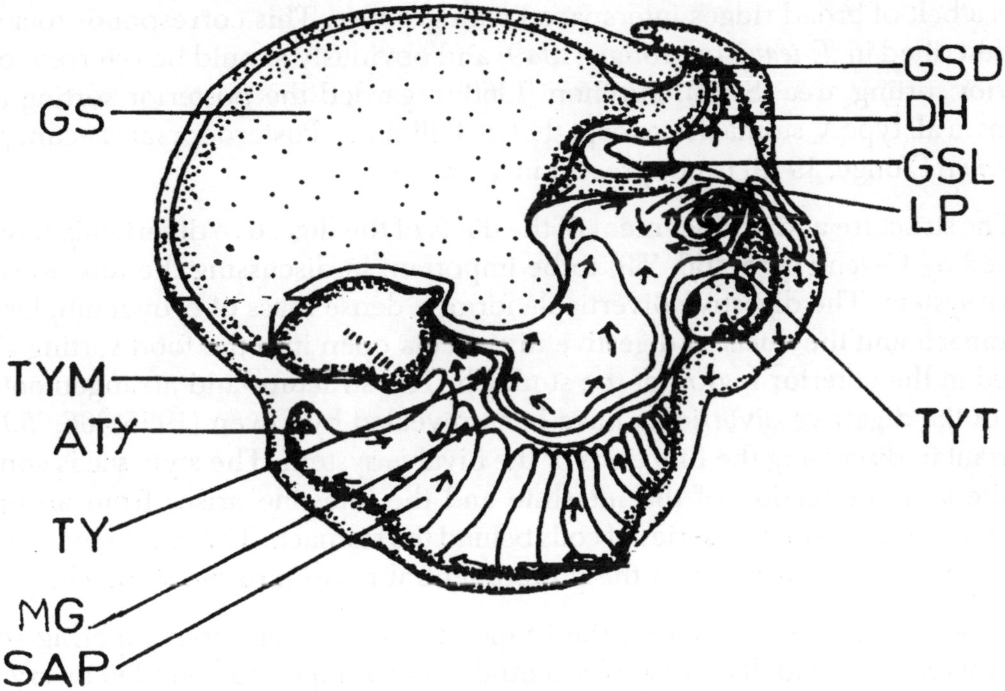

Fig. 10 : Diagram showing stomach of *T. ala*.

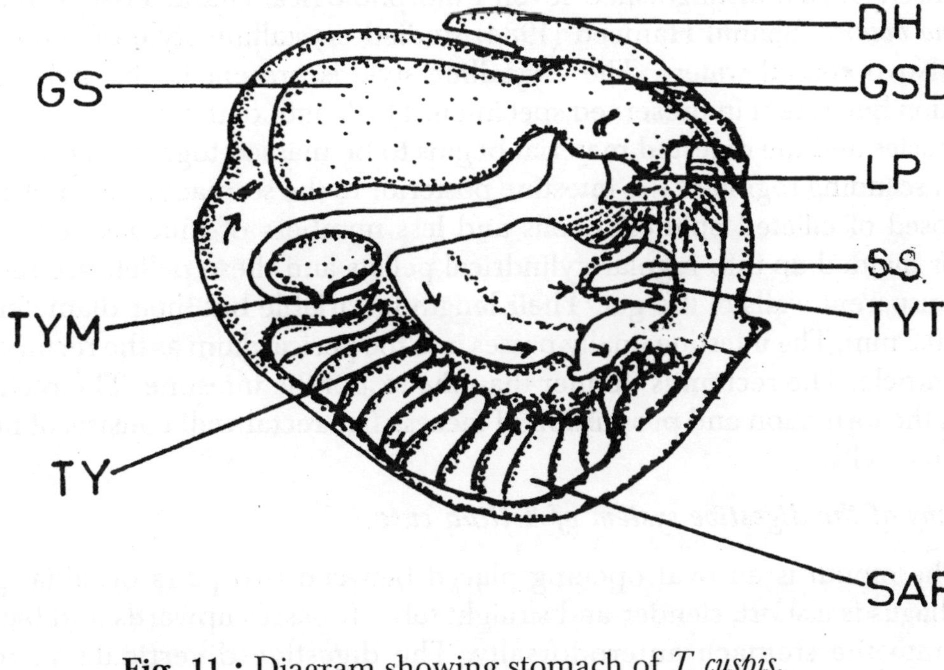

Fig. 11 : Diagram showing stomach of *T. cuspis*.

wall is a belt of broad ridges interspaced with grooves. This corresponds to a similar area described in *T. tenuis* by Yonge (1949) and obviously should be referred to as the posterior sorting area (SAR). Purchon (I960) regarded the posterior sorting area as present in all type V stomachs, except that of Tellinidae. Posterodorsal caecum present in *T. tenuis* (Yonge, 1949) is not present in *T. ala*.

The structure and arrangement of the ducts of the digestive diverticula have been revealed by Owen (1955, '66, '59) to be important in discussing the functions of the bivalve system. The digestive diverticula forms a dense mass of brown tubules below the stomach and the ducts of digestive diverticula open into the food sorting caecum situated in the anterior region of the stomach. The structure and arrangement of the ducts of the digestive diverticula have been revealed by Owen (1955, '66, '59) to be important in discussing the functions of the bivalve system. The style sac is conjoined with the anterior section of the intestine and the intestine arises from an opening situated ventrally forming a series of coils behind the stomach. The combined crystalline style sac and midgut arise from the posteroventral portion of the stomach.

In general the style assumes the shape of a long gelationous tapering rod and curved at the distal end. It consists of a central granular liquid surrounded by crystalline coaxial layers (Pl. 4). The morphology of the bivalve crystalline style, has been examined by Nelson (1918, '25), Edmondson (1920) and Yonge (1923, '26b, '49). Kristensen (1972a) distinguished seven I morphological characteristics in the style of *Macoma bathica*. Shahul Hameed (1984) studied crystalline style of some bivalve of Porto Novo coastal waters. The crystalline style is present in the style sac in fresh condition but absent in preserved specimen. The length of the intestine varies in these two species and the digested material begins to be massed together about the middle of the ascending region of the intestine posterior to the stomach. The intestinal wall is composed of ciliated epithelial cells and less numbers of mucous cells. The initial mass is divided up into regular cylindrical pellets and these pellets are seen through the transparent walls of the gut. Their length is variable but their diameter is usually about 0.2 mm. The intestine finally passes into the pericardium as the rectum traversing the ventricle. The rectum is thinner than the rest of the intestine. The rectum plays a role in the formation and propulsion of faeces. The rectal wall consists of non-ciliated columnar cells.

Anatomy of the digestive system of Tellina cuspis

The mouth is an oval opening placed between two pairs of labial palps. The oesophagus is a short, slender and straight tube. It passes upwards and backwards to open into the stomach anterodorsally. The digestive diverticula surrounds the oesophagus and the anterior part of the stomach. The combined style sac and midgut leaves from the posteroventral wall of the stomach and runs downwards to the base of the foot. The midgut takes a forward turn leaving a small pocket of style sac, completely separated from the midgut. The midgut undergoes coiling immediately below the stomach and emerges as the hindgut. It dips downwards and follows a course along the periphery of the visceral mass and runs through the pericardium and the heart. It opens out through the anus at the posterior adductor muscle.

3.4. REPRODUCTIVE SYSTEM

In both species the sexes are separate and the gonads occur as a massive unpaired structure. The gonads are well developed encircling the intestine and imbricated in the digestive gland. The testis and ovary consists of many follicles in which the germ cells are seen. Small channels leading from these follicles unite to form paired genital ducts which open into the mantle cavity through the urinogenital papilla. During maturation the follicles are smaller in size, gradually occurring more space and spread over digestive diverticula. The oocytes start increasing in size by the accumulation of yolk in the cytoplasm (Pl. 19). Spermatozoa are arranged in lamellae and the tails are directed towards the centre of lumen, çocytes are compactly arranged and attached to the wall of the follicle (Pl. 23 and 24).

4

BREEDING BIOLOGY

4.1. INTRODUCTION

Adequate knowledge on the breeding biology of animal is essential in predicting annual recruitment and to interpret growth, especially for practicing successfully the culture of commercially important bivalves. This in turn requires an understanding of spawning behavior and reproductive cycles of molluscs. Reproductive cycles of various commercially important bivalves have been studied from Indian waters as well as in other waters.

Reproductive cycle and sexual changes in clams have been studied from foreign waters by Loosanaff and Davis (1950) on genus *Mercenaria*. Reproductive cycle and spawning in the soft shell clam *Mya arenaria* were studied by Stickney (1952), Rogers (1959) and Brusseau (1978). Ansell (1961) reported the reproductive cycle and spawning season of the clam *Venus striatula*. Reproduction and growth in *Tellina fabula* were reported by Guy Bachelet (1980). He also investigated the reproductive cycle and recruitment of Tellinid bivalve *Macoma balthica*.

There are many works on the reproductive biology and gonadial cycle in bivalves from Indian waters. Some of the important works are by Hornell (1922), Rai (1932), Abraham (1953), Durve (1964), Salih (1973), Harkantra (1975), Sreenivasan (1980) and Balasubramanian (1984) on *Meretrix casta*; Marimuthu (1985) studied the reproductive cycle in *Crassostrea madrasensis* and *Mytilus edulis;* Rao (1967), Mane (1974), and Kalyanasundaram (1982) observed sexual maturity and sexual cycles in *Katelysia opima*; Alagarswami (1966) on *Donax faba*; Nayar (1955), and Rao (1967) on *Donax cuneatus*; Narasimham (1968) on *Anadara granosa,* George John (1980) on *Anadara rhombea,* Dravidamani (1985) on *Sunetta scripta* and Hornell (1922), Rai (1932), Nagabhushanam and Deshmukh (1974) and Jayabal (1984) on *Meretrix meretrix*. The breeding biology of Tellinids has not so far been studied from Porto Novo waters and hence the present study was undertaken in Vellar estuary to fill up the existing lacuna.

4.2. MATERIALS AND METHODS

Specimens of *T. ala* and *T. cuspis* ranging (50 to 100 number) were collected monthly from the marine zone of the Vellar estuary and their reproductive condition was recorded based on the extent to which the gonads spread over the digestive diverticula. Fresh gonad smears were also examined under a microscope to ascertain the sex and maturity of the gonad. For histological studies, pieces of gonad tissues were fixed in Bouin's fixative, embedded in paraffin wax and sections were cut at 6 to 8 u thickness. Ehrlich's haematoxylin and eosin were used for staining the sections. Different maturity stages were classified as described by George John (1980).

4.3. OBSERVATIONS

Histological and smear observations of gonads were made from June 1987 to May 1988 to determine the sex, stage and size of maturity. Breeding season of Tellinids was inferred from these observations.

4.3.1. SIZE AT FIRST MATURITY

In the present investigations it was found by smear observations that majority of *T. ala* attain first sexual, maturity at 11 to 13 mm in size (length) and *T. cuspis* at 8 to 11 mm in size. Sometimes the immature Tellinids could be recorded even upto 15 mm and 13 mm in size in *T. ala* and *T. cuspis* respectively (Table 5, 6, 7 and 8). Males and females attained sexual maturity at the same time.

4.3.1. SEX RATIO

The sex ratio of the population was calculated every month from June 1987 to May 1988; Chi-square test was applied to test whether the population follows 1:1 ratio (Table 9). From the total number of specimens observed during the present study, the female and male ratio was found to be 1:0.91 in *T. ala* and 1:0.84 in T. *cuspis* showing slight dominance of females in both species. Chi-square analysis revealed significant deviation from 1:1 ratio only during two months (April and May, 1988). In all other months the ratio was nearly 1:1 with slight predominance of female in the population.

The sex ratio of different size groups of Tellinids are shown in Table 9 and 10. Females were dominant in size groups 11-13 mm, 14-16 mm, 17-19 mm, and 26-28 mm in *T. ala* and males were dominant in 20-22 and 23-25 mm. Whereas in *T. cuspis* females were dominant in all size groups examined. In 26-28 mm (*T. ala*) and 24-26 mm (*T. cuspis*) length groups the ratio was exactly 1:1. In Chi-square analysis, sex ratio confirmed the expected 1:1 ratio for all the size groups.

Based on the development of gonad, specimens of *T. ala* and *T. cuspis* could be categorized into 3 groups as follows.

(a) *Immature*

Below 11 mm (in length) in *T. ala* and 8 mm in *T. cuspis* the specimens showed no sign of sexual maturity. In this group differentiation of sex was not possible as no trace of male or female germ cells could be seen.

(b) *Mature Males and Females*

In *T. ala* and *T. cuspis* in specimens above 16 and 13 mm respectively the male and female gonads are well developed showing various stages of maturity in smear examinations.

(c) *Spent*

After spawning there is no trace of sexuality. The number and percentage occurrence of males, females and spent in both species are presented in Table 8. Even though males and females occur throughout the year, immature forms did not dominate during monsoon and postmonsoon months (October to March). From April onwards the number of young ones increased since fresh recruitment to the populations started. Spent animals could be recorded from January to October in both species. Males and females were divided into two groups maturing and mature stage.

4.3.3. MATURITY STAGES

In both species of *Tellina,* sexes are separate and the gonad (male and female) consists of numerous round or oval follicles, which increase in size during maturation. Follicles are branched and the ducts which collect ova or spermatozoa are situated on the sides and each duct connected with many follicles. All these ducts join to form the primary genital duct which opens into the mantle cavity through the genital aperture. During maturation, the follicles are smaller in size. They gradually occupy more space and spread over digestive diverticula. The follicle walls are thin with a layer of spermatogonia. Spermatocytes and spermatids are closer to the periphery of the follicles. The oocytes start increasing in size by the accumulation of yolk in the cytoplasm. In the mature stage the follicles are fully spread, covering the digestive diverticula extending even up umbo region. Spermatozoa are arranged in lamellae and their tails are directed towards the centre of the lumen. Oocytes are compactly arranged and attached to the wall of the follicles. Spawning starts when most of the ova in the follicles become detached and are found free in the lumen. The gonad has been classified into maturing and mature based on the smear observations and sections of the gonads.

Maturing in *Tellinids*

The percentage of maturing stage is high in September (34.18 in male and 34.99 in female) in *T. ala* and in *T. cuspis* 31.5 in female and 30.62 in male. During maturation the follicles are smaller in size and gradually occupy more space and spread over the digestive diverticula. The oocytes start increasing in size by the accumulation of yolk in the cytoplasm. Spermatocytes and spermatids are closer to the periphery of the follicles.

Mature

In this stage the follicles are fully spread covering the digestive diverticula. Oocytes are compactly arranged and attached to the wall of the follicles. Spawning starts when most of the ova in the follicles became detached and are found free in the lumen. High parentage of mature stage in both species was observed in the months of February to March. The percentage of maturing and mature stages of both male and female is presented in Table 7 and 8. Both males and females of Tellinids attain full maturity at the same time. It could be inferred that in both species there is no variation in breeding season. And both males and females mature at the same time, and it could be observed that Tellinids breeds for an extended period from February to September.

4.4. DISCUSSION

The present study reveals that *T. ala* attains first sexual maturity at 11-13 mm and 8-10 mm in *T. cuspis*. Rao (1951) and Mane (1974) observed that the size at first maturity of male and female in *Katelysia opima* was 11.4-12.0 mm and 12-13 mm in length respectively. Kalyanasundaram (1982) found *K. opima* to attain maturity at 24.7 mm size. George John (1980) found that *Anadara rhombea* attained first maturity at 21-25 mm size in Porto Novo waters and Jayabal (1984) reported that *Meretrix meretrix* attains sexual maturity at 21-26 mm in size.

Studies on the sex ratio of *T. ala* and *T. cuspis* in the present study shows that

Table 5. Number and percentage of male, female and spent *T. ala.* (above 8 mm in length).

Months	*T. ala*						*Total*
	Female		*Male*		*Spent*		*Number*
	N	*%*	*N*	*%*	*N*	*%*	
June 1987	10	23.80	9	21.43	23	54.76	42
July	14	36.84	10	26.31	14	36.84	38
August	15	39.47	13	34.21	10	26.31	38
September	18	47.36	12	31.58	8	21.05	35
October	17	48.60	17	48.57	1	2.86	38
November	16	53.33	14	46.67	0	0.00	30
December	18	47.36	20	52.63	0	0.00	38
January 1988	17	50.00	16	47.06	1	2.94	34
February	18	48.65	16	43.24	3	8.11	37
March	19	50.00	15	39.47	4	10.53	38
April	12	32.43	16	43.24	9	24.32	37
May 1988	14	32.56	15	34.88	14	32.56	43

Table 6. Number and percentage of male, female and spent *T. cuspis* (11 mm in length).

Months	*Female*		*Male*		*Spent*		*Total*
	N	*%*	*N*	*%*	*N*	*%*	*Number*
June 1987	5	18.52	10	37.08	12	44.44	27
July	7	36.84	5	26.31	7	36.84	19
August	8	40.00	7	35.00	5	25.00	20
September	9	47.37	6	31.58	4	21.05	19
October	9	47.37	9	47.37	1	5.26	19
November	8	53.33	7	46.67	0	0.00	15
December	9	47.37	10	52.63	0	0.00	19
January 1988	9	50.00	8	44.44	1	5.55	18
February	9	50.00	8	44.44	1	5.55	18
March	10	50.00	8	40.00	2	10.00	20
April	6	31.58	8	42.10	5	26.31	19
May 1988	7	31.82	8	36.36	7	31.82	22

Table 7. Monthly percentage distribution of the maturity stages of male and female in *T. ala*

Months	*Female*		*Male*	
	Maturing %	*Mature* %	*Maturing* %	*Mature* %
June 1987	12.33	37.66	12.66	37.33
July	24.13	25.87	23.71	26.28
August	31.60	18.39	32.15	17.84
September	34.99	10.81	34.18	15.30
October	25.38	24.62	26.42	23.57
November	24.29	25.70	21.03	28.97
December	23.51	26.48	19.87	30.63
January 1988	10.62	39.38	9.36	40.13
February	14.50	35.50	0.57	49.43
March	–	100.00	–	100.00
April	5.01	44.99	0.53	49.47
May 1988	6.16	43.83	6.22	43.77

Table 8. Monthly percentage distribution of the maturity stages of male and female in *T. cuspis*

Months	*Female*		*Male*	
	Maturing %	*Mature* %	*Maturing* %	*Mature* %
June 1987	9.11	32.33	7.33	32.00
July	19.13	20.43	19.34	21.28
August	30.30	13.14	28.72	12.42
September	31.50	5.40	30.62	10.30
October	20.14	19.50	21.42	15.57
November	19.14	20.35	16.03	23.97
December	18.25	21.48	14.43	25.63
January 1988	5.30	34.19	4.18	35.50
February	5.50	49.50	0.50	44.40
March	–	100.00	–	100.00
April	10.00	39.50	0.50	50.00
May 1988	8.00	21.44	3.11	22.54

Table 9. Sex ratios and Chi-square values for *T. ala* in different size groups

Size groups (mm)	*Female* (n)	*Male* (n)	*Female : Male*	*Chi-square* value		*P*
11 – 13	103	94	1 : 0.912	0.394	>	0.05
14 – 16	165	160	1 : 0.969	0.076	>	0.05
17 – 19	147	119	1 : 0.809	2.666	>	0.05
20 – 22	28	32	1 : 1.142	0.286	>	0.05
23 – 25	28	31	1 : 1.107	0.160	>	0.05
26 – 28	1	1	1 : 1.000	0.000	>	0.05

Table. 10. Sex ratios and Chi-square values for *T. cuspis* in different size groups

Size groups (mm)	*Female (n)*	*Male (n)*	*Female : Male*	*Chi-square value*		*P*
8-10	10	7	1 : 0.700	0.450	>	0.05
11-13	26	21	1 : 0.807	0.480	>	0.05
14-16	32	28	1.142 : 1	0.286	>	0.05
17-19	52	46	1.130 : 1	0.392	>	0.05
20-22	58	46	1.260 : 1	1.566	>	0.05
23-25	1	1	1 : 1.000	0.000	>	0.05

Table 11. Sex ratios and Chi-square values during various months in Tellinids (*T. ala.* and *T. cuspis*)

	T.ala. 643 specimens		*T. cuspis* 667 specimens	
Months	F : M	X^2	F : M	X^2
June 1987	1 : 0.92	0.08	1 : 0.81	0.60
July	1 : 0.94	0.22	1 : 0.91	1.00
August	1 : 0.78	0.72	1 : 0.87	0.26
September	1 : 0.92	0.07	1 : 0.85	0.29
October	1 : 0.85	0.32	1 : 0.83	0.40
November	1 : 0.71	1.66	1 : 0.92	0.07
December	1 : 0.57	4.26*	1 : 0.78	0.42
January 1988	1 : 1.50	1.00	1 : 0.79	0.66
February	1 : 0.92	0.07	1 : 1.03	0.01
March	1 : 0.76	1.06	1 : 0.85	0.32
April	1 : 1.17	0.32	1 : 0.57	4.26*
May 1988	1 : 0.83	0.45	1 : 0.62	3.26*

F = Female; M = Male ;

P = Value in all cases > 0.05 except in * where $P < 0.05$

females are slightly dominant in numbers. The sex ratio of male and females of Tellinids fluctuated monthly and also in different size groups but the chi-square analysis showed that the sex ratio is confirming the expected 1:1 ratio, monthwise, (except in two months) size-wise when the data were pooled for the whole year. Some important works on breeding biology of bivalves like *Mytilus edulis aoteanus, Aulacomya maorina* (Kennedy, 1977); *Mya arenaria* (Keck *et al*., 1975) reveal that the temperature influence spawning in temperate water. However, in tropics salinity rather than temperature influence spawning. Durve (1964) reported that very low salinity may have some adverse effects on the spawning activity of bivalves. So during monsoon time, there is no spawning. Similar observations are made by Kalyanasundaram (1982) in *Crassostrea madrasensis.* He reported that spawning was initiated by the rise in salinity and temperature. He observed that throughout the monsoon period, these mature oysters remained in the same condition due to the prevailing low salinity and after monsoon the oysters started to spawn.

Rao (1951) reported from the Adayar estuary at Madras on the east coast that *Katelysia opima* spawns twice in a year. Sreenivasan (1980), George John (1980), Kalyanasundaram (1982) and Jayabal (1984) studied the spawning seasons of *M. casta, A. rhombea, K. opima* and *M. meretrix* respectively from Porto Novo waters and these species have prolonged spawning season. Similar observations are noted in the present study also. Tellinids have breeding season from February to September. It was also reported in other trophical animals *M. Casta (*Abraham, 1953) and in *Donax Faba* (Alagarswami, 1966).

Guy Bachelet (1980) reported in *Macoma balthica* that gonad development started in June, as soon as spawning was completed. Recovery of condition continued throughout the summer months, so that mature gametes were again present in late October. He also reported that analysis of data for individual years revealed some variations from one year to another. Ansell and Body (1979) observed gonadial development in April/May and spawning in August/December in *Donax trunculus.* They observed that in *D. trunculus,* reproductive period continued from April to September. Kalyanasundaram (1982) reported that *K. Opima* from east coast of India, spawned from April to May. Ansell (1967) observed in Tellinids, gonad development takes place during early summer and spawning in June. He opined that reproductive period may be in the summer when the temperature rises above a certain limiting level, and they growth, gonad growth, and spawning proceed together in response to environmental changes like the availability of food and occurrence of specific spawning stimuli, etc. Abraham (1953) found that *M. casta* from Adayar estuary spawns actively from July to August, October to November and March to April. He also suggested that the spawning period does not remain constant and show year to year variation in the same environment. *Donax cuneatus* has a single spawning season as reported by Nayar (1955). Alagarswami (1966) found that *Donax faba* has spawning period from Feb. to Sep. in a year. In the present study maturation takes place round the year with a peak in February and March.

The present study reveals that Tellinids have prolonged breeding seasons from February to September in the Vellar estuary. In both species, male and female attain first sexual maturity at the same time. Females are slightly dominant than male in the population. Salinity seems to play an important role and the reproductive cycle starts form March onwards and no breeding of Tellinids was observed during monsoon period.

5

SEASONAL VARIATIONS IN BIOCHEMICAL CONSTITUENTS OF *Tellina ala* and *Tellina cuspis*

5.1. INTRODUCTION

Tellinids are important as key members of the fauna of sandy shores and possibly as major items in the diet of young fish in such areas where experiments directed to the development of fish rearing on a commercial scale are being pursued (Ansell and Trevallion, 1967). Marine molluscs lack the descrete biochemical storage sites like adipose tissue as in other animals hence the storage occurs in the somatic and/or reproductive organs through the production of new cellular elements (Giese, 1959). The storage of biochemical nutrients is essential in bivalves since these nutrients are utilized during the active reproductive period whether the environmental conditions are favourable or unfavourable to reproduction (Giese *et al.*, 1967). Seasonal changes in the major nutrient reserves of the soft tissues have been extensively studied by Ansell and Trevallion (1967) in *Tellina tenuis;* by Galtsoff (1964) in *Crassostrea viriginica;* by Ansell (1974) in *Nucula salcata* by Giese *et al* (1967) in *Tivela stultorum;* by George John (1980) in *Anadara rhombea;* by Jayabal (1986) in *Meretrix meretrix*; and by Balasubramanian (1984) in *Meretrix casta.* Biochemical constituents related to reproductive cycle in the different body components of various *bivalves on* a seasonal basis were studied by Galtsoff, (1964), Owen (1966), Ansell (1967, '74), Giese, Hart and Smith (1967), Saraswathy and Nair (1969) and George John (1980). Investigation on biochemical composition in different body parts, of an animal would be more informative than estimation of the same in the entire body for studies related to seasonal variations (Giese, 1969). Studies on biochemical composition in different soft tissues of various bivalves on a seasonal basis were studied by Owen (1966), Ansell (1967, '74), Saraswathy, and Nair (1969); George John (1980) and Jayabal (1984).

In the present investigation an attempt was made to ascertain variations in nutrients reserves of Tellinids with respect to seasonal changes and reproductive cycles as the information is not available in the literature.

5.2. MATERIALS AND METHODS

Adult Tellinids (*T. ala* and *T. cuspis)* were collected monthly for a period of one year from Vellar mouth of the estuary from June 87 to May 88. Animals were brought to the laboratory and kept in filtered estuarine water for one day. The total length of these animals were measured using Vernier callipers and adult Tellinids of the same size group (25 to 30 mm) were selected for analysis. Males and females were separated by smear observations. The soft body parts of both sexes were removed after a quick rinse with distilled water and were oven dried for 24 h at 60°C. The different body parts namely mantle, gonad, digestive diverticula and adductor muscle from 20 males and 20 females were excised for analysis and the soft body parts were ovendried for 24 h at 60°C and then finally powdered for estimating biochemical indices. Protein was estimated by Biuret method as described by Raymont *et al.,* (1964). Method of Dubois *et al.,* (1956) was followed for the determination of total carbohydrate content. The estimation of lipid was done by employing the chloroform methanol method of Folch *et al.,* (1956) and then analysed in the Hitachi double beam spectrophotometer (model 12208). Protein was read at 540 nm and carbohydrate at 490 nm. The values are expressed as percentage of dry weight. Five estimations were made in each case and the average was taken.

5.3. RESULTS

5.3.1. PROTEIN

WHOLE ANIMAL : Protein values for the whole animal and for different body components of both sexes of Tellinids (*T. ala* and *T.cuspis*) are shown in (Fig. 36, 37, 38 and 39). The percentage of protein in male of *T. ala* exhibited fluctuations from 18 to 28% and in female from 23% to 32%. In *T. cuspis* the range of protein in male varied from 14 to 22% and in female from 19 to 28%. The highest value was noticed in April while the lowest in September. The percentage of protein in both sexes of Tellinids increased gradually through October to May and recorded a peak in April and this may be due to onset of maturation. Whereas reduction in percentage of protein was observed in September due to active spawning. Generally the biochemical constituents in both sexes of Tellinids were high during maturation period (February - April) and lower during breeding season (September to November).

DIFFERENT BODY PARTS

The range of protein values in different body parts of *T. ala* and *T. cuspis* are given below.

Protein in the gonadal tissue of Tellinids varied as follows:

	Male	*Female*
T. ala	19 to 25%	22 to 32%
T. cuspis	17 to 22%	18 to 25%

The percentage of protein in the gonads of Tellinids were observed to be higher during maturation period (April) than during September, i.e. spawning period. Generally in the gonad percentage of chemical constituents were maximum during reproductive period.

DIGESTIVE DIVERTICULA

It was observed that the lowest percentage of biochemical constituents was recorded during reproductive period in digestive diverticula. Highest percentage was recorded in September and the lowest in April.

Protein values determined in the digestive diverticula of Tellinids are as follows :

	Male	*Female*
T. ala	18 to 24%	22 to 32%
T. cuspis	18 to 22%	19 to 31%

MANTLE

The highest value was recorded in May and the lowest in December.

	Male	*Female*
T. ala	12 to 18%	14 to 19%
T. cuspis	10 to 11%	12 to 15%

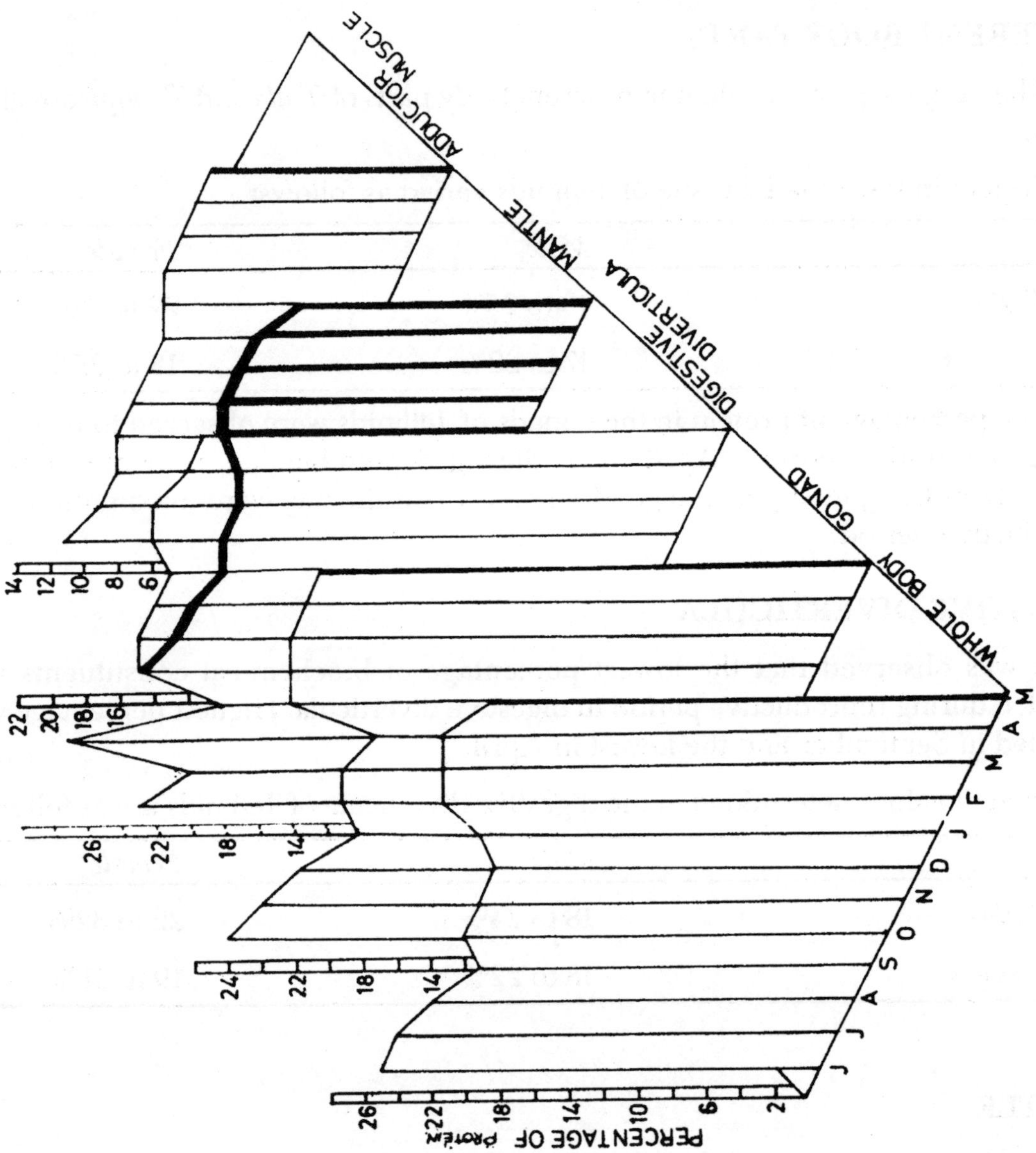

Fig. 12 : Monthly protein variations in whole animal and different body components (Gonad, Digestive diverticula, Mantle and Adductor muscle) of female of *T. ala.*

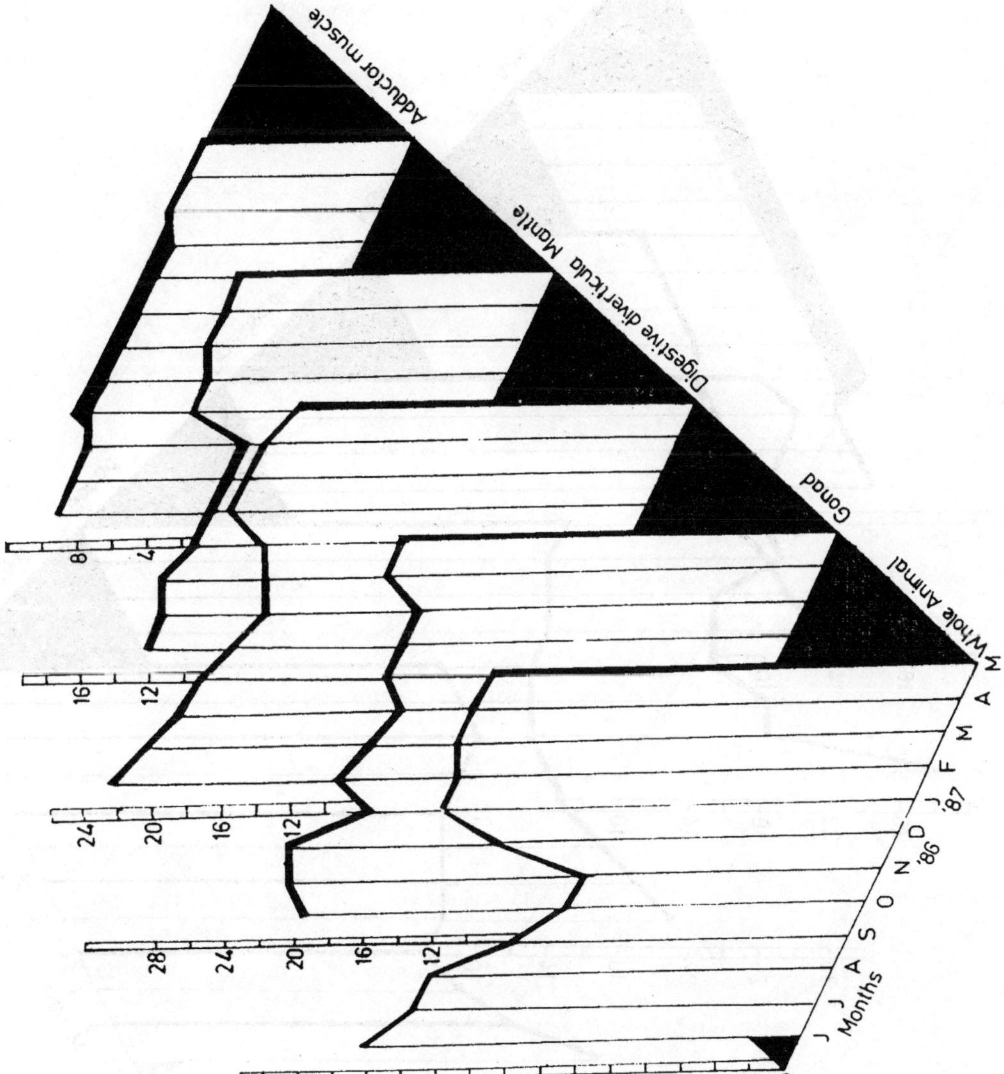

Fig. 13 : Monthly protein variations in whole animal and different body components of male of *T. ala*.

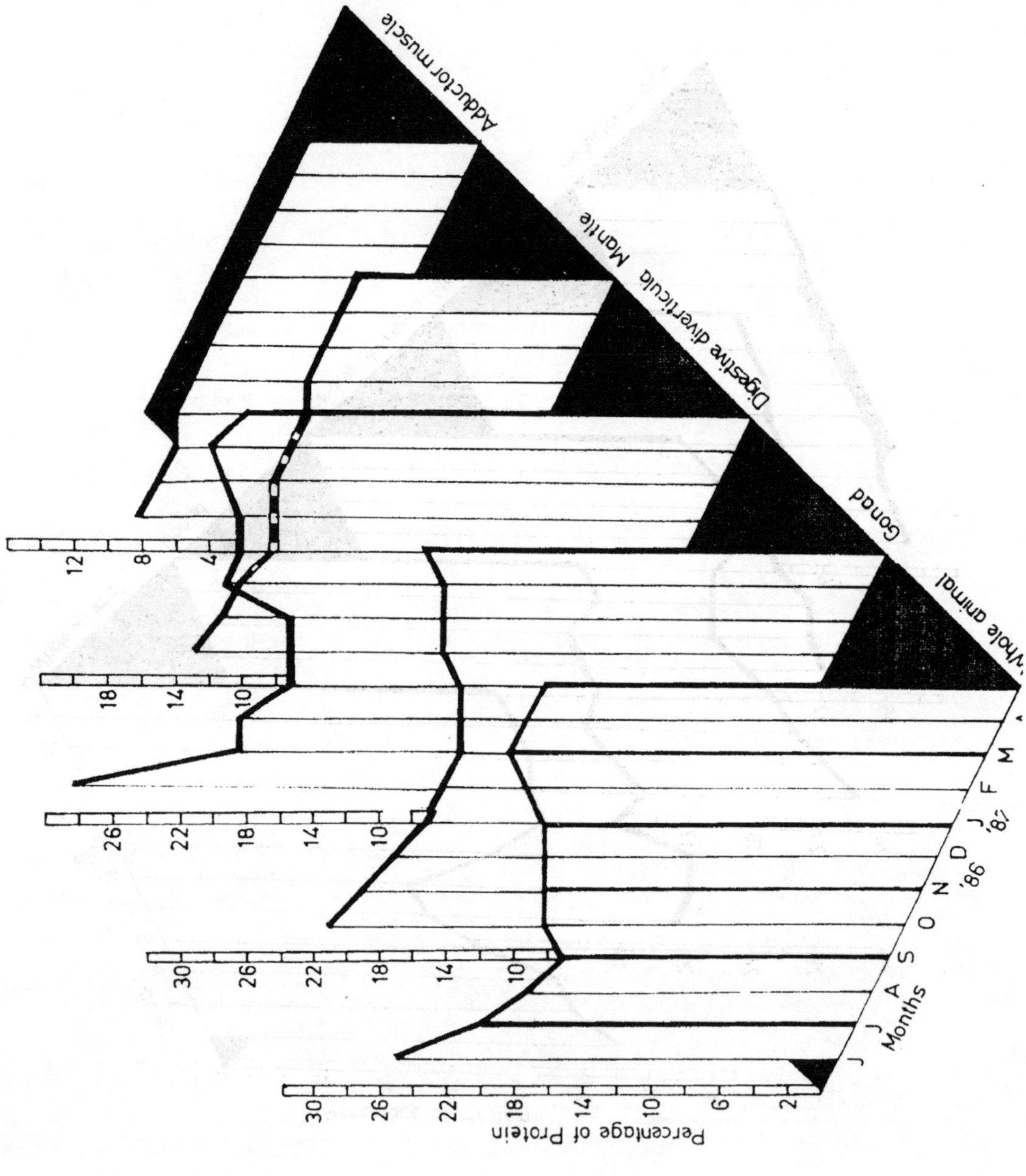

Fig. 14 : Seasonal variations in protein in whole animal and different soft tissues of female of *T. cuspis.*

ADDUCTOR MUSCLE

The range of protien values in the adductor muscle of Tellinids are given below.

	Male	*Female*
T. ala	11 to 12%	12 to 14%
T. cuspis	9 to 10%	9 to 10%

Least fluctuation in the percentage of biochemical constituents of adductor muscle were observed.

5.3.2. CARRBOHYDRATE

Fluctuations in carbohydrate content in whole and different body parts of both sexes of Tellinids are given below. Monthly variations are given in (Fig. 16 to 19)

Maximum percentage of carbohydrate was observed in April and minimum in September. It gradually increases through the monsoon months upto April. It fluctuates 3.9 to 4.2% in male and 4.5 to 7.8% in female of *T. ala* and in *T. cuspis* the values ranged between 3.5 and 4 in male and in female from 4 to 6.2%.

	Male	*Female*
T. ala		
Whole animal	3.9 to 4.2%	4.5% to 7.8%
Gonad	8.9 to 10.5%	9.9 to 12.1%
Digestive diverticula	7.8 to 9.9%	8.3 to 11.2%
Mantle	6.5 to 7.1%	7.1 to 7.5%
Adductor	4.5 to 5.0%	4 to 5%
T. cuspis		
whole animal	3.5 to 4.0%	4.0% to 6.2%
Gonad	7. 2 to 9.1%	7. 3 to 9.5%
Digestive diverticula	6.9 to 7.8%	7.2 to 7.9%
Mantle	4.4 to 5.1%	4.6 to 5.2%
Adductor	3.9 to 4.0%	4.2 to 4.5%

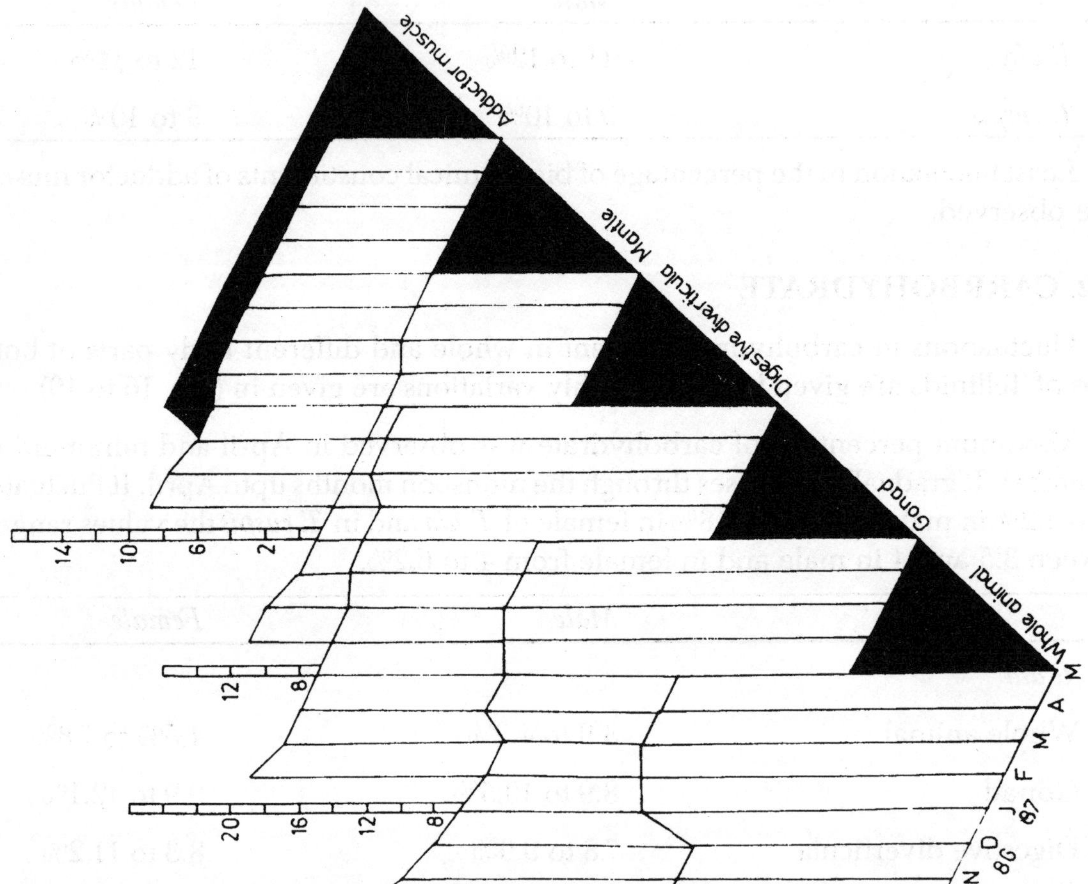

Fig. 15 : Monthly protein variations in whole animal and different body components of male *T. cuspis.*

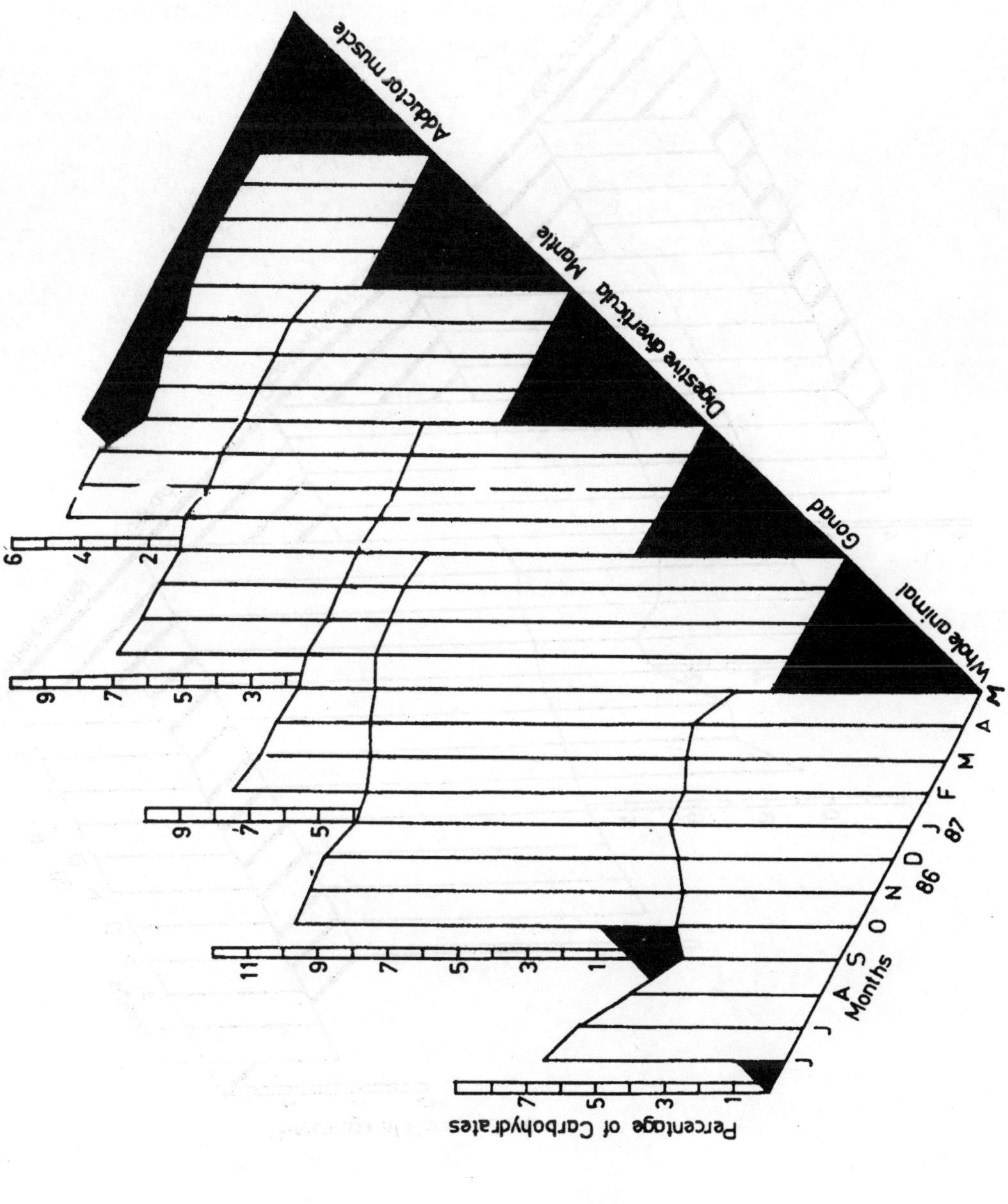

Fig. 16 : Diagram showing monthly carbohydrate variation in whole animal and different body components of female of *T. ala.*

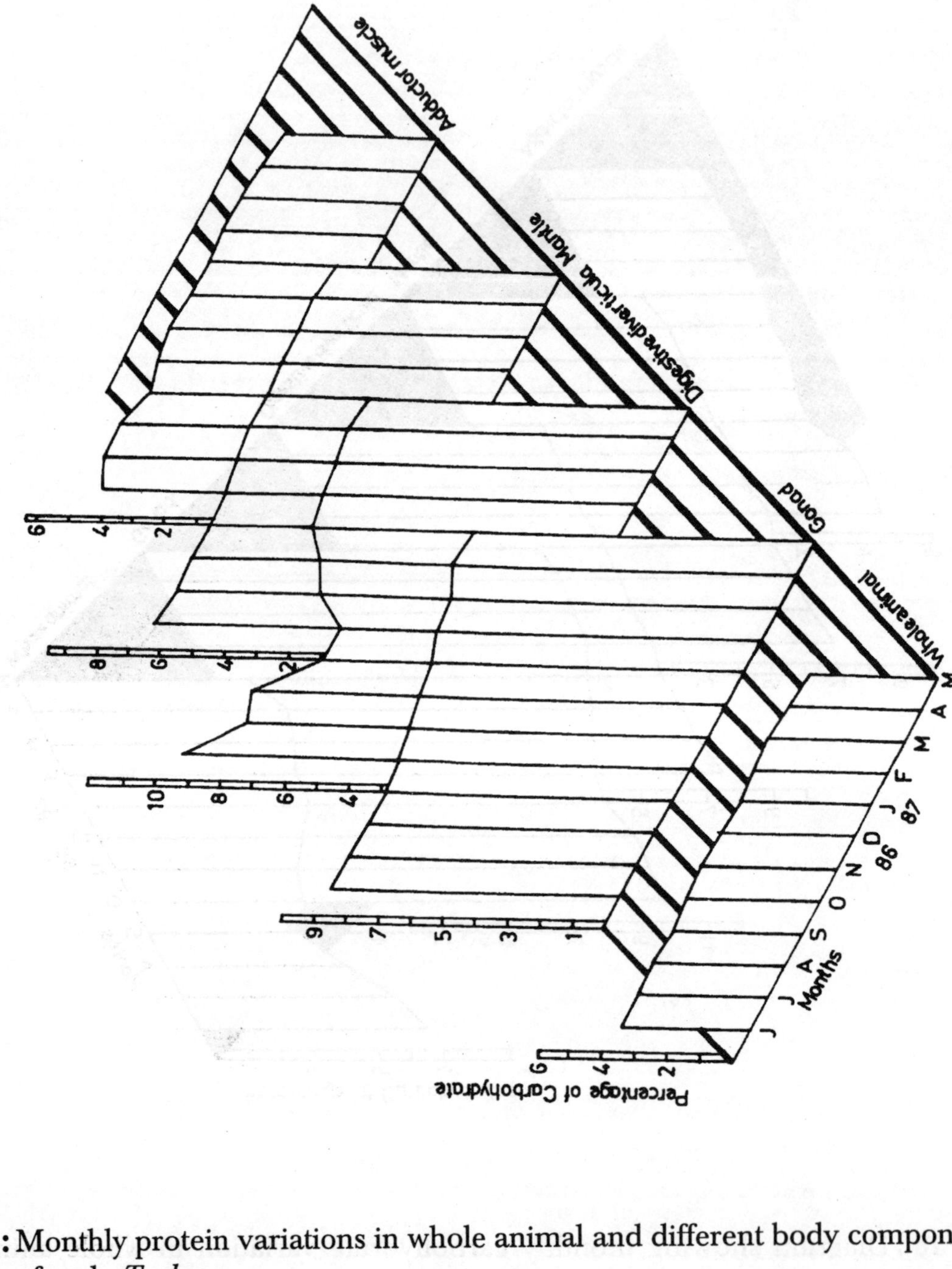

Fig. 17 : Monthly protein variations in whole animal and different body components of male *T. ala*.

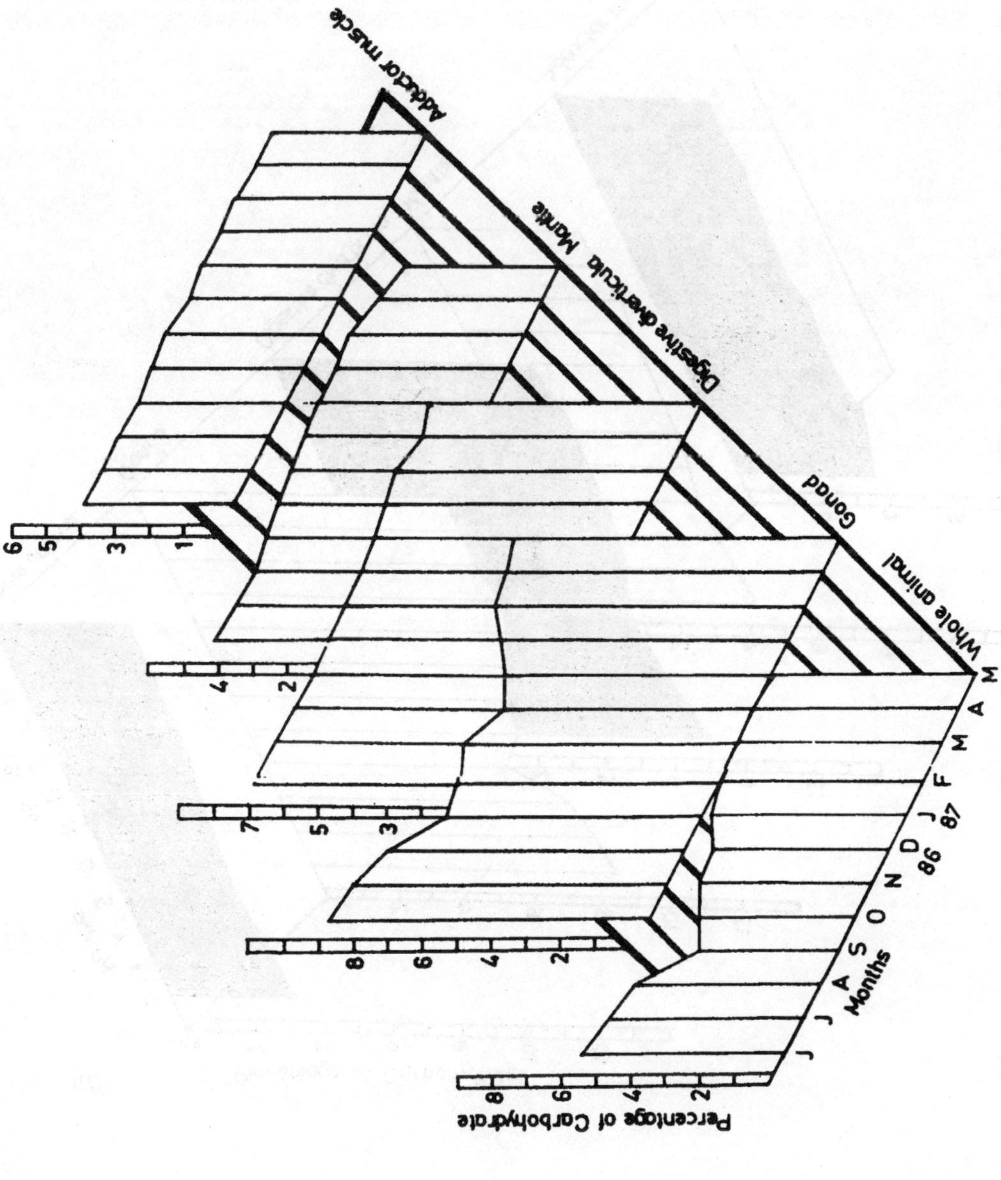

Fig. 18 : Monthly carbohydrate variations in whole animal and different body components of Male of *T. cuspis.*

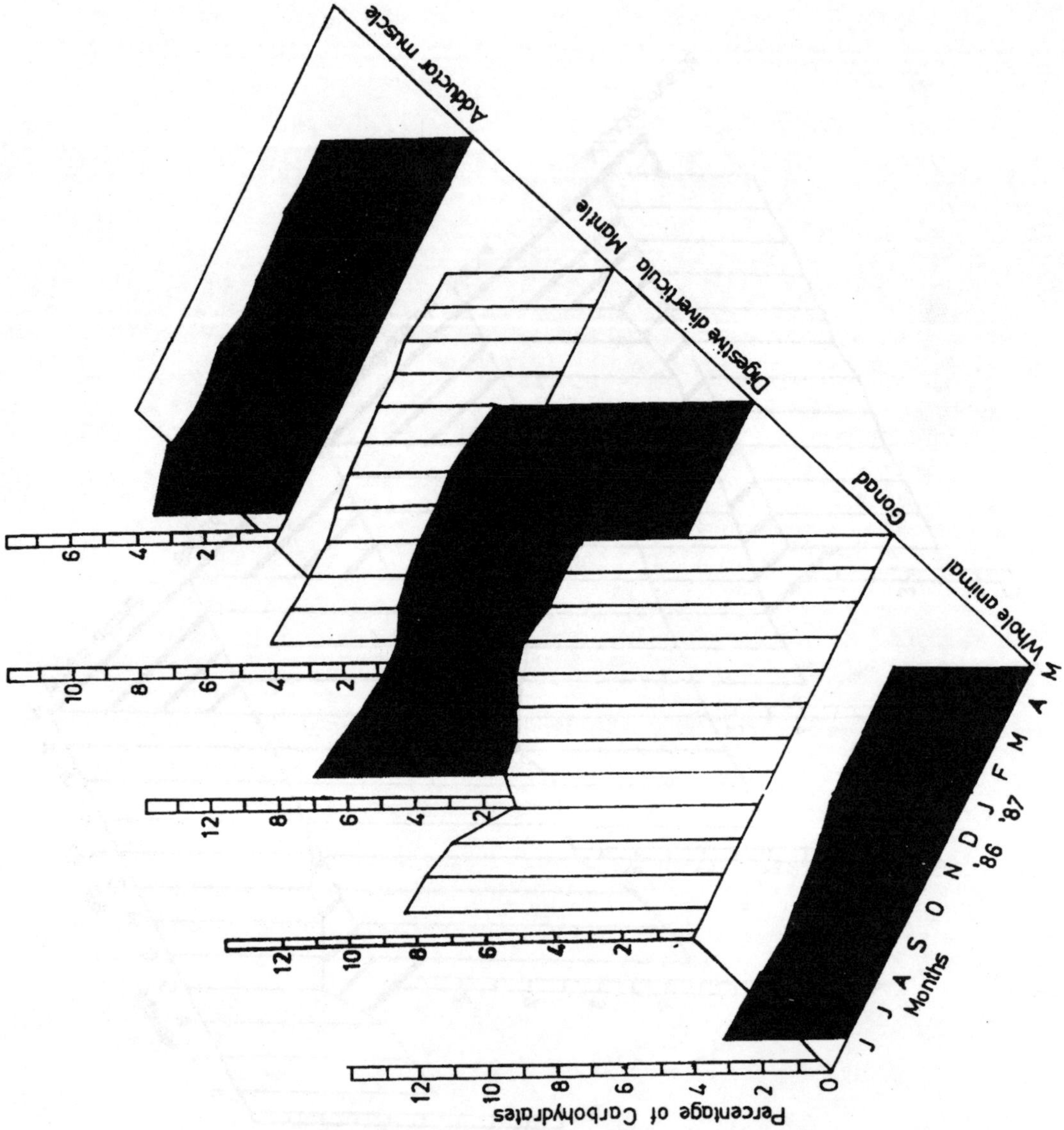

Fig. 19 : Monthly carbohydrate variations in whole animal and different body components of male of *T. cuspis.*

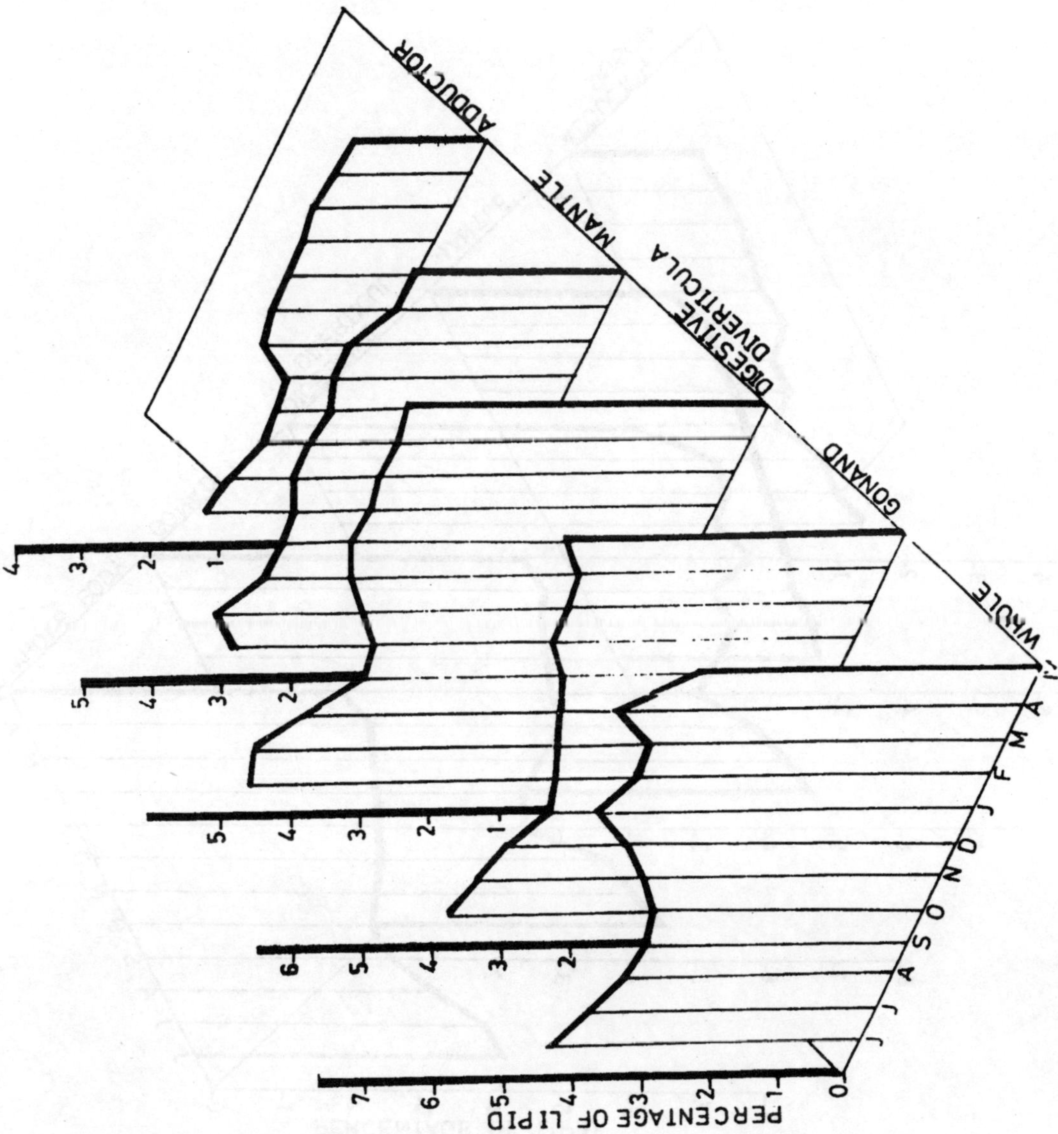

Fig. 20 : Montly lipid variations in whole animal and different body components of female of *T. ala*.

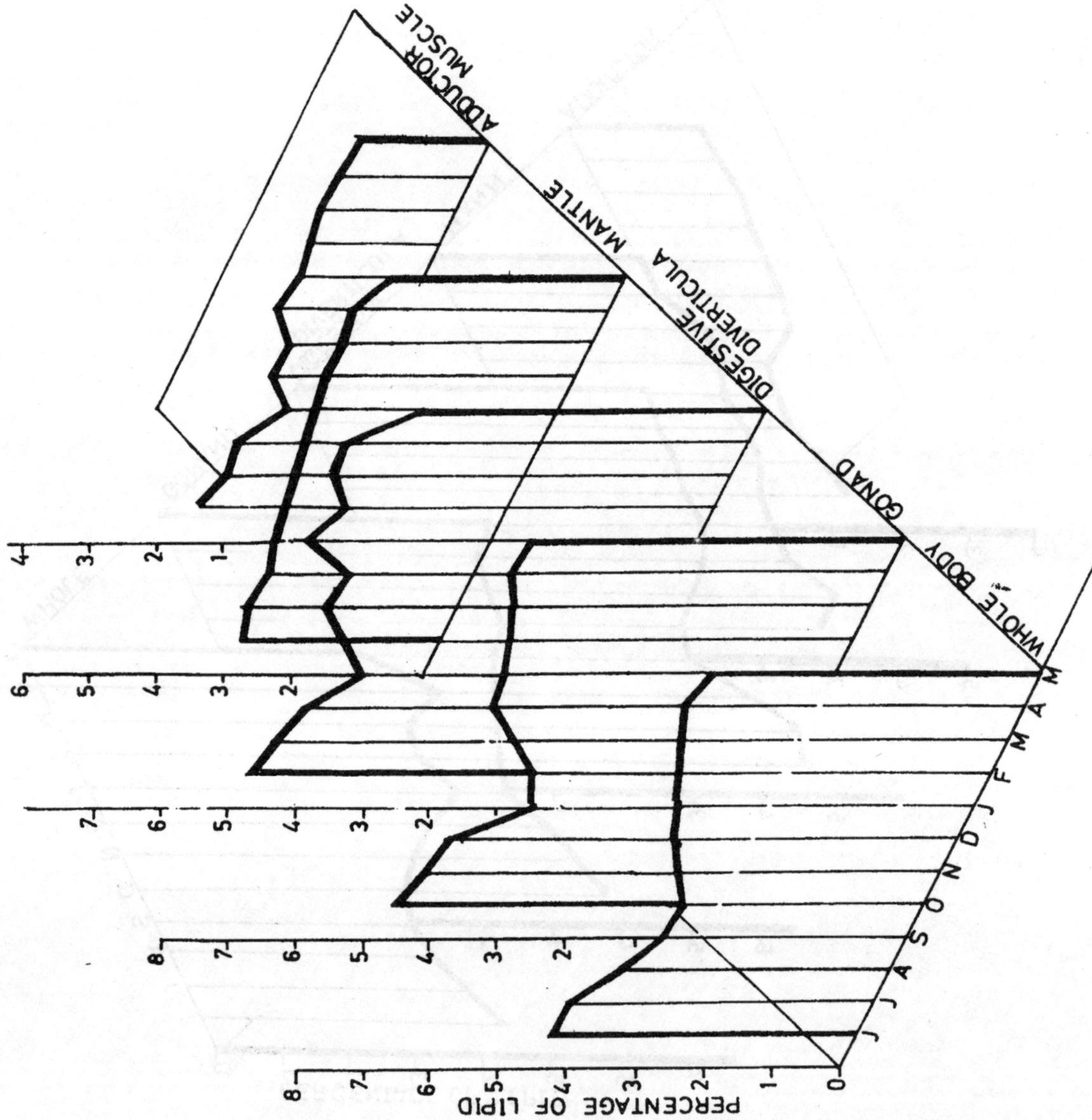

Fig. 21 : Monthly variations in lipid of whole animal and different body components of male of *T. ala.*

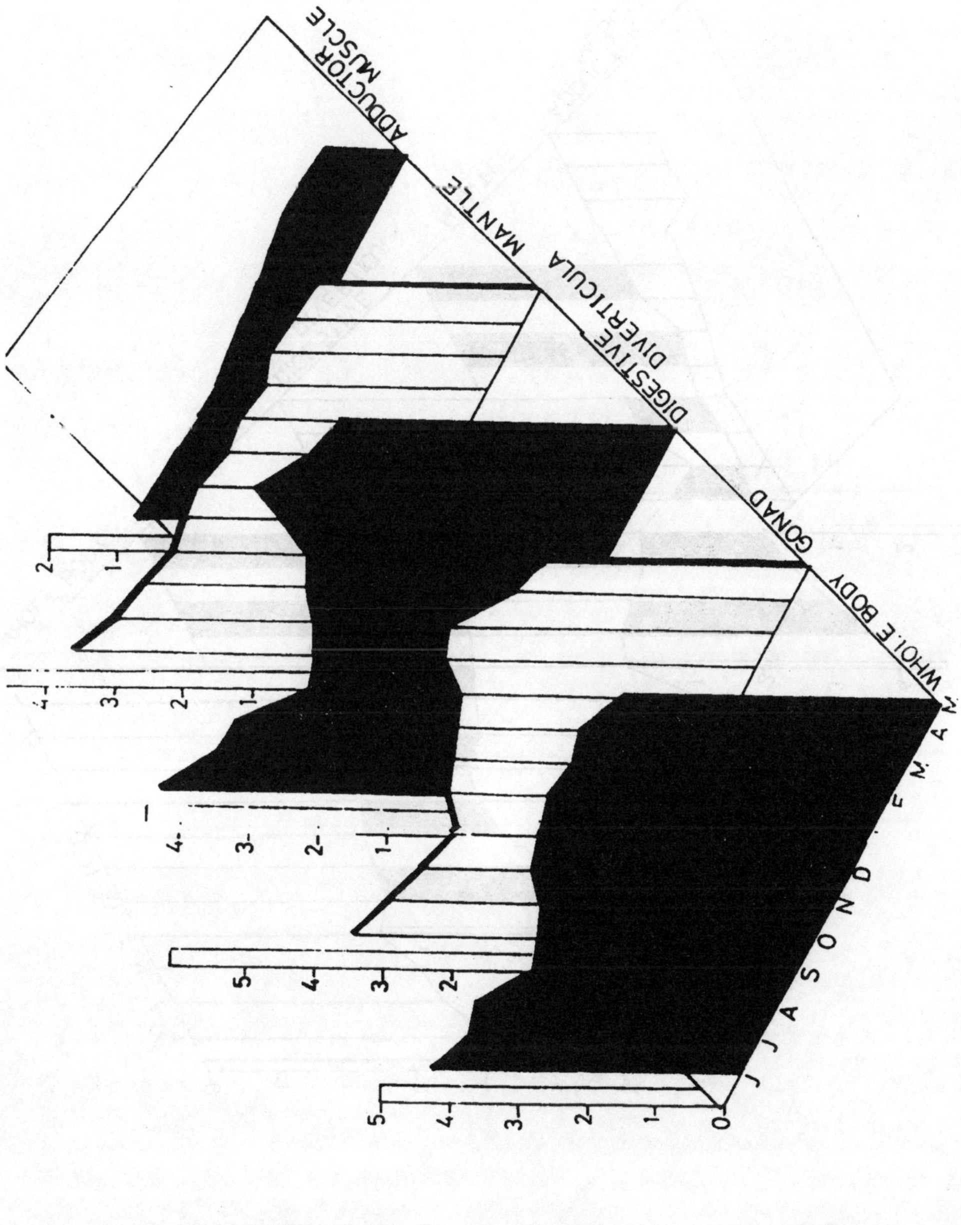

Fig. 22 : Seasonal variations in lipid of whole animal and different body components of female of *T. cuspis.*

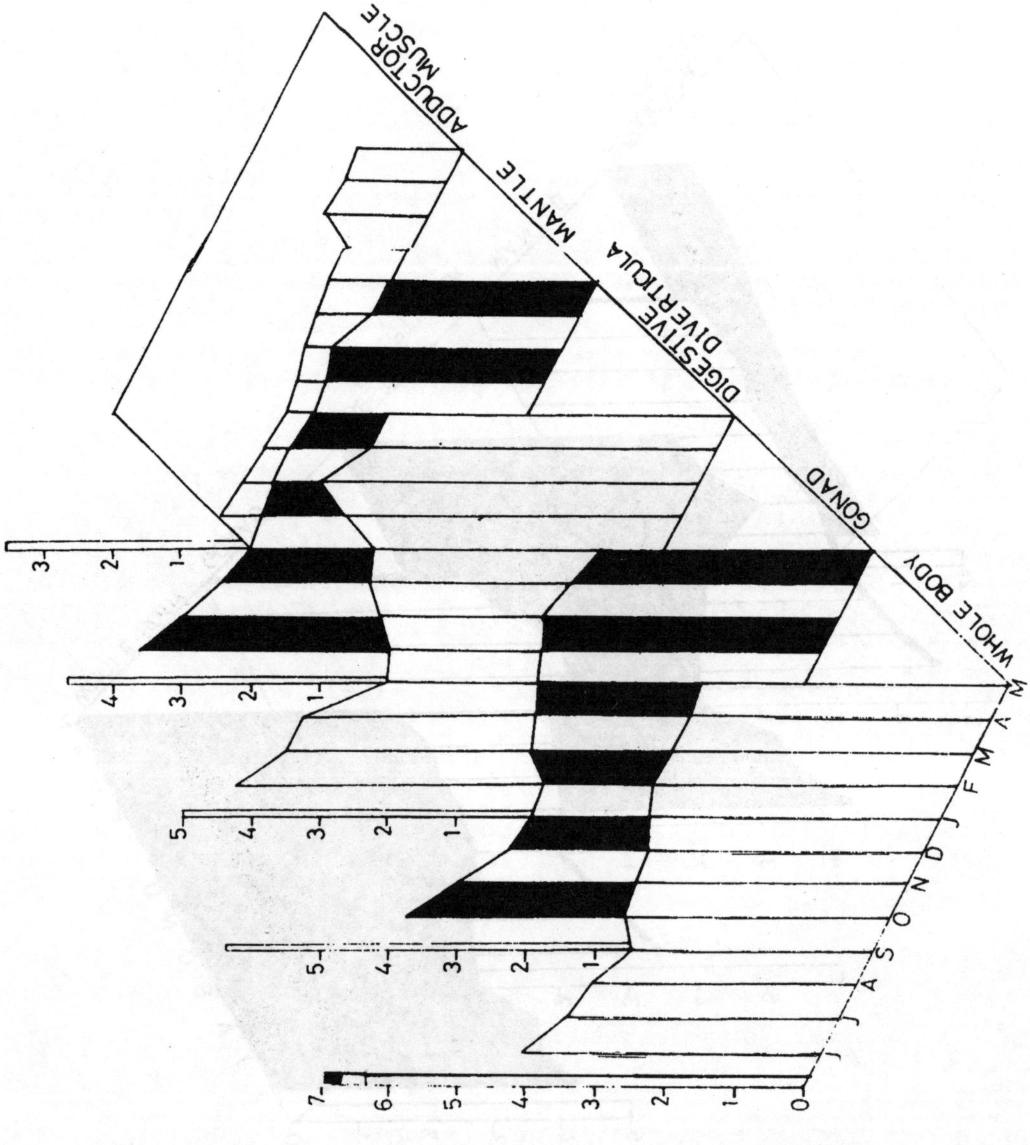

Fig. 23 : Monthly variations in whole animal and different body components of male of *T. cuspis*.

5.3.3. LIPID

Generally the fat content in the whole organism and different soft tissues of Tellinids were low. The highest value was observed in April and lowest values during September. Digestive diverticula showed the maximum content of lipids as compared to other parts of the body (Fig. 20 to 23).

	Male	*Female*
T. ala		
Whole animal	3.6 to 5.0%	3.8 to 5.2%
Gonad	3.5 to 5.5%	3.2 to 4.9%
Digestive diverticula	4.0 to 6.0%	3.9 to 5.2%
Mantle	3.3 to 3.9%	3.0 to 3.5%
Adductor muscle	1.0 to 2.0%	1.0 to 1.5%
T. cuspis		
Whole animal	3.5 to 4.5%	3.8 to 5.0%
Gonad	2.9 to 4.2%	3.0 to 4.7%
Digestive diverticula	3.0 to 5.5%	3.1 to 5.5%
Mantle	3.0 to 3.3%	3.2 to 3.4%
Adductor muscle	0.5 to 1.0%	0.8 to 1.2%

5.4. DISCUSSION

From the results of the present study, it is apparently inferred that the biochemical constituents were highest during the maturation period and lowest during spawning period. Variations in biochemical indices seem to be mainly influenced by reproductive cycle (George John, 1980). The present study indicated that fluctuations in the biochemical constituents studied presently is related to maturation of gonads and the reproduction cycle. In the present study the observations on the state of development of the gonads showed the first signs of gonad development occurring in April. Increase in biochemical constituents occurring between March and June is associated with the proliferation of gonadal tissue in the visceral mass. And it was also observed that in the gonad and digestive diverticula the chemical indices cycled inversely.

Gonad indices were maximum during the reproductive period whereas the digestive diverticula indices were minimum during this period. Similar observations were studied in *Meretrix meretrix* by Nagabhushanam and Deshmukh (1974) and in *Meretrix meretrix* by Jayabal (1984). From the present observations it was also found that the biochemical constituents are slightly higher in females than in males as reported in *Anadara rhombea* (George John, 1980).

Least fluctuations in the chemical constituents of mantle and adductor muscle were observed. The nutrients stored in somatic parts were utilized during active spawning period. A period of inactivity during the winter months, during which gametogenesis may proceed slowly and when resources, particularly of glycogen stored in various tissues may be drawn up to supply reduced metabolic demands, a short period following the renewal of activity in the spring. When reserves are renewed, growth recommences and rapid gonad proliferation and gametogenesis occur. A productive period during the summer when the temperature rises above a certain limiting body growth, gonad growth and spawning proceed together in response to environmental changes in, for example, the availability of food and occurrence of specific spawning stimuli. In their seasonal biochemical cycle, *T. ala* and *T. cuspis* resembles *T. tenuis* and the clam *Merceneria mercenaria* (Ansell *et al.*, 1964; Ansell and Lander, 1967). Protein is the main constituent followed by lipid and carbohydrate. The percentage of protein, carbohydrate and lipid values increased from February to April due to the maturation of gonads and later declining trend was observed up to September mainly due to active spawning. In Tellinids, immediately after spawning, maturation of gonads takes place and, therefore, biochemical constituents slowly increase through monsoon months (October-December) up to March. This is very similar with the observation made by Ansell (1974) in *Tellina tenuis* and George John (1980) in *Anadara rhombea.*

Among different body fractions, digestive diverticula recorded high values of protein, carbohydrate and lipid except during active spawning period. The digestive diverticula acts as a storage site in most of the bivalves (Owen, 1966; George John, 1980). A similar observation was also noted in the present study that variations in the chemical indices of digestive diverticula indicate that this organ acts as a possible storage site to Tellinids and the adductor and mantle to lesser extent. It was observed that the chemical constituents cycled inversely in the gonad and digestive diverticula. It was also noticed that during reproductive period the percentage of chemical constituents of the digestive diverticula was lowest. In mantle during dry season (February-May) when there is active feeding and synthesis, the protein values were more. In general all the biochemical constituents (Protein, carbohydrate and lipid) in both sexes of

Tellinids (*T. ala* and *T. cuspis)* showed maximum content during maturation period and minimum during spawning period in both whole body and different soft body tissues. The chemical constituents of gonad was maximum during reproductive period and minimum in digestive gland. The percentage of chemical constituents in gonad and digestive diverticula cycled inversely.

Least fluctuations in the chemical constituents of mantle and adductor muscle were observed. The percentage of chemical constituents in *T. ala* was greater than *T. cuspis.* Similarly the percentage of chemical constituents were high in female than the male. A general increase in biochemical constituents was observed summer months and extremely low reserves during monsoon is supported by the occurrence of a high mortality rate among Tellina at Firemore during the winter of 1965-66 (McIntyre, 1968). From November to March the biochemical constituents decrease slowly indicating utilization of reserves by the animal during the monsoon months. From the present study it is concluded that seasonal variations in biochemical constituents in the whole body and different body fractions of Tellinids were mainly influenced by reproductive cycle.

6

PESTICIDES EFFECTS ON TELLINIDS

6.1. GENERAL INTRODUCTION

The discharge of effluents waste materials from factories, oilspills from vessels, widespread and indiscriminate use of the pesticides in agriculture and also domestic sewage have been causing a great concern as the environment is being dangerously contaminated. Among the organic chemical pollutants, pesticides are the chief sources of contamination and are highly toxic. Klein (1962), in a bid to classify water pollutants, kept the chemical pollutants in the first rank of which pesticides are the major source of contamination.

Synthetic pesticides have, in recent years, become increasingly important additions to chemical wastes polluting natural aquatic communities. They have been introduced for more than 40 years ago. Ever since the discovery of the insecticidal action of DDT by Paul Muller in 1946, the use of pesticides has increased rapidly (Rajendran, 1984). Widespread and extensive uses of pesticides in recent years has led to grave concern, over the consequences of such usage especially on non-target organisms. Pesticides are used extensively on a worldwide basis in forestry, agriculture and in public health to control a wide variety of insect pests (Sundaramoorthy, 1987).

Pesticides applied to crops, forests and for public health are generally spread from the site of their application to estuaries and seas by direct drift as well as through surface run off (Tarrant and Tatton, 1968; Murray and Vaugham, 1969; Frost and Ware, 1970; Frere, 1971). Accidental spilling, washing of agro-chemical devices and disposal of pesticide containers in streams and rivers and discharge of industrial and agricultural waste waters from other sources of pollution of natural waters by pesticides

(Rajendran, 1984). As much as 40-60% of the pesticides used on land do not reach the target organisms, but find their way into estuarine and marine environments through river discharge, terrestrial runoff, industrial, agricultural and municipal drainage and by atmospheric transportation (GESAMP, 1976).

Of the wide range of pesticides currently in use, those of serious concern to organisms are organochlorines rather than organophosphates and carbamates (Sundaramoorthy, 1987). Organochlorines are known to be highly persistant and toxic to large numbers of target organisms (Rajendran, 1984). Because of their greater persistence, low solubility in water, strong adsorbance to particulate matter, and their potential to bioaccumulate, the organochlorine pesticides are considered highly toxic and known to adversely affect the aquatic living resources.

Concentrations of organochlorines in the seas around India have been reported to be high than in other areas (Tanabe and Tatsukawa, 1980). Greve and Wit (1971) reported that endosulfan was present at a concentration of 70 ug/1 in water sampled from the Rhine estuary after a massive fish kill. Fairly high amounts of organochlorine pesticide residues have been recorded in Vellar estuary along with information on their toxicity to various marine organisms (Rajendran, 1984). Compounds such as DDT, Lindane and endosulfan are commonly used organochlorine pesticides in India.

Due to their nonpersistence and quick degradability, organophosphates and carbamates do not pose a serious threat to aquatic life in general and estuarine and marine life in particular. But organochlorines have practically proved to be a menace due to their long persistence and nondegradability (Sasibhushana Rao, 1980).

Pesticides are potential pollutants that pose greatest threat to marine organisms. Pesticides enter the biogeochemical cycles operating in the marine environment and cause deleterious effects in its resident organisms (Sundaramoorthy, 1987). Many of these pesticides are considered hazardous because of their adverse effects on fresh water, estuarine and marine organisms even at extremely low concentrations (Tarzwell, 1963; Cope,.1965; Eisler, 1977). While generally not causing direct mortality they may well have sublethal effects reducing the health and viability of marine organisms (Roberts, 1975). Certain organochloride insecticides like DDT and endosulfan retain their toxicity for months or even years (Bridges *et al.*, 1963; Woodwell and Wurster, 1967). Organophosphate insecticides are known to cause hyperglucaemia and also can alter the biochemical constituents of the organisms (Rajendran, 1984). Organophosphorus biocides degrade rapidly in biological system releasing the metabolic products into irrigation and drainage canals, thus affecting aquatic fauna (Rao, 1980). The translocation of pesticides from agricultural fields to the aquatic environment poses a serious threat to fishes and shellfishes of commercial importance. Shellfishes are considered a group of organisms that can usefully be utilized as test animals for pesticides in the environment (Butler, 1969). Shellfish have the capacity to take up pesticides directly from the water in which case the entry site is primarily through the gills and secondarily through the gut (Brodtmann, 1970). The route of entry of pesticides is generally agreed to be via the gills (Holden, 1962; Ferguson *et al.*, 1966) and hence the respiratory process may be adversely affected.

Among shellfishes, bivalves received much importance and are used as indicator species due to their sedentary nature. Being filterfeeder they seem to accumulate the pesticides to several times even from a very low concentration in the ambient medium. Tellinids are small benthic deposit feeding bivalves. At present, they are consumed by a limited number of people, and are also used in pisciculture practices as feed to fishes (Ansell, 1967)

Considerable data, based on bioassay have been generated concerning the effects of organochlorine pesticides on shell growth of oyster and mortality of clams (Waugh and Ansell, 1956; Loosanoff, 1947; Engel et al. 1972; Schimmel et al. 1977a). A review of literature has showed that physiological activities like feeding, respiratory mechanisms, excretion and various life activities such as ciliary activity, burrowing and byssal activity have been seriously impaired in fishes and shellfishes on exposure to pesticides even in low concentrations. Muley and Mane (1987) studied histopathological changes induced by Cythion-malathion in the gonads of lamellibranch molluscs. Ciliary activity in fresh water mussel *Lamellidens marginalis* exposed to pesticide in various temperatures were reported by Basha and Swami (1987). Reddy *et al.*, (1988) reported effect of phosphamidon and DDT on respiratory activity of the Penaeid prawns *Metapenaeus monoceros.*

Increasing attention is being paid to the sublethal effects of pollutants on marine organisms, to obtain baseline data which would be useful in assessing the impact of pollution. The importance of measuring changes in behaviour, physiology and metabolism has been recognised as essential to the assessment of chronic effects of low levels of contaminants in ocean. The use of such tests as routine bioassays was suggested by Stirling (1975) who reported effects on the burrowing behaviour of the bivalve *Tellina tenuis* exposed to sublethal concentrations of copper and phenol.

Though there are several reports on pesticide toxicity to fresh water fauna in India (Rao, 1980; Verma *et al.* 1975; Santharam *et al.* 1976) reports on marine or estuarine organisms are very limited (Subramanian *et al.* 1979; Lingaraja *et al.* 1979). And hence in the present study in addition to biological studies, effects of pesticides (endosulfan and dimethoate) on physiological and histological changes of tellinids (*T. ala* and *T. cuspis)* were also studied.

6.2. MATERIALS AND METHODS

6.2.1. MATERIALS

The adult animals were collected from the marine zone of the Vellar estuary and were brought to the laboratory, cleaned and acclimated in well aerated water for three days at 28°C before the start of experiments. The animals ranging from 30 to 32 mm *(T. ala)* and 16 to 18 mm (*T. cuspis)* shell length were selected for the experiments. The test organisms were fed with fresh phytoplankton.

6.2.2. WATER

Estuarine water was pumped and stored in the overhead tank of aquarium. It was filtered through glass wool column and well aerated before use. Salinity 28 ± 0.2 ‰o was maintained throughout the experiment. All the parameters such as salinity (28 ± 0.2‰o), temperature (28 ± 0.5°C), dissolved oxygen (4.5 ± 0.2 ml/1) and pH (8 ± 0.2) were regularly monitored during the experiment.

6.2.3. CONTAINERS

45 1 capacity clean, all fibre glass rectangular tanks of 30×30×50 cm dimensions were used in both acute and sublethal toxicity studies.

6.2.4. PREPARATION OF PESTICIDES CONCENTRATION

The experiments were performed using an organochlorine pesticide. Endosulfan (Technical grade; purity 96%) the product of Bharat Pulversing Mills Private LTD., Bombay and organophosphorus pesticide Dimethoate (Technical grade; 94% purity) obtained from Rallis India Ltd., Bombay-400 001 (Table 12). Acetone (AR grade) was used as a carrier sovent to prepare stock solutions of the pesticides which has no effect on the activity of animals up to a level of 0.5 ml/1. The pesticide was dissolved in appropriate volume of acetone to obtain the stock solution. The appropriate test concentrations were obtained by adding the stock solution to the test aquaria. From the stock solution different concentrations were prepared.

Table 12

Pesticide	*Supplied by*	*Technical Grade*	*Chemical Name*
Endosulfan (Organo-chlorine)	Bharat Pulversing Mills Private Ltd. Bombay	96%	6, 7, 8, 9, 10 – Hexa-chloro – 1, 5, 5a, 6, 9, 9a hexahydro – 6, 9, methano – 2, 4, 3, benzo dioxathiepin 3 – oxide $C_9H_6Cl_6O_3$
Dimethoate (Organo-phosphate)	Rallis India Ltd. Bombay-400 001	94%	0, 0 – Dimethyl S- (N-methyl carbomoyl-methyl) phosphorodi-thioate

Table 13

Pesticides	Concentration of Pesticides (ppb) *T. ala.*				*T. cuspis.*			
	LC_{50}	SL_1	SL_2	SL_3	LC_{50}	SL_1	SL_2	SL_3
Endosulfan	10	4	2	1	6	3	2	1
Dimethoate	15	8	4	2	10	5	3	4

6.2.5. METHODS

Acute toxicity tests were 96 hr static but renewable type of bioassay conducted to estimate the 96 hr LC50 values. The test media were renewed for every 24 hrs. No feed was given during the experimental period. Ten randomly distributed organisms were tested at each concentration. Dead bivalves were removed from the experimental tanks immediately. Two sets of controls were maintained, with and without added solvent. The bivalves were considered dead when they kept open their shell valves even after gentle prodding. The mortality of the animals in each test concentration of both pesticides was recorded at every 6 hr interval for 96 hrs. No mortality occurred in control and solvent treated bivalves. Litchfield and Wilcoxon (1949) method was employed to calculate the LC50 with 95% confidence limits. Based on the LC50 values three sublethal concentrations were selected (Table 13).

6.3. PESTICIDE TOXICITY ON RESPIRATION OF TELLINIDS

6.3.1. INTRODUCTION

Pesticides, because of their potential toxicity, are known to produce morphological, behavioural and physiological changes to different animals (Fingerman, 1982). Respiration is frequently used as an indicator of general condition or stress. Any change in respiration rate may indicate detrimental physiological damage which may be just as important as mortality (Thurberg *et al* 1980). To screen for possible effects of pollution, respiration studies have successfully been used invitro (Thurberg, 1980).

The system which gets affected in living organisms due to the toxic effects varies with metals, pesticides and toxin produced by animals and the effects extend from interference with enzymatic activities to impairment of vital physiological functions such as respiration and osmoregulation (Sunil Kumar Mohamed, 1980). High concentrations of heavy metals, pesticides and organic wastes are found in coastal waters. Further, organisms *in coastal* waters are also subjected to fluctuating environmental conditions. Coastal and estuarine waters are heavily polluted as a result of continuing urbanization and industrialization. Sewage effluents, industrial wastes and pesticides from agricultural fields are the main sources for environmental pollution. Metal effects on respiration have been extensively studied, since oxygen consumption can be used to asses the physiological stress of the animal (Brown and Newell, 1972). The rate of oxygen consumption, especially in poikilothermic organisms, may change rapidly and is affected by many endogenous and environmental factors (Newell, 1979). Oxygen consumption is a valuable indicator of energy expended to meet the demands of an environmental alteration (Thurberg *et al* 1975). They reported the effect of silver on respiratory response of larval, juvenile and adult surf clams, *Spisula solidissima.* They also examined the effects of silver on the gill tissue and oxygen consumption in four bivalve species, *Mercenaria. mercenaria, Crossostrea, virginica, Mytilus edulis* and *Mya arenaria.* Although there are several reports available on the physiological effects of pollutants in marine animals only a few of these concern molluscs (Eisler and Gardner, 1973). In view of the inadequate information, the present study was undertaken to investigate the extent of pesticide effects on the respiratory physiology of Tellinids.

6.3.2. MATERIALS AND METHOD

Tellinids of uniform size group ranging from 30 to 32 mm (*T. ala*) and 24 to 26 mm (*T. cuspis)* were used in the present study. The biological conditions of the experimental animals, collection, acclimatisation and preparation of the pesticide test concentrations, etc., are described in the earlier chapter (6.2).

The 96 hr LC_{50} value was found out as described earlier in section (6.2). Based on the LC_{50} value, two sublethal concentrations were chosen individually for the two species as mentioned in the previous chapter (Table 13).

A continuous flow system designed by Lingaraja *et al* (1979) and later modified by Rao (1980) was set up to estimate the oxygen consumption in *T. ala* exposed to lethal (10 ppb) and sublethal (4 and 2 ppb) concentrations of endosulfan, and in *T. cuspis* exposed to lethal (6 ppb) and sublethal (3 and 2 ppb) concentrations of endosulfan. Similarly *T. ala* and *T. cuspis* were also exposed to acute and sublethal concentrations (15, 8 and 4 ppb; 10, 5 and 3 ppb) of dimethoate respectively for a period of 96hrs. Control was also run simultaneously.

Six animals were introduced into each experimental chamber individually and were left undisturbed in the respiratory chamber for a period of 2 hrs prior to the commencement of the experiment so as to allow the animals to get acclimatized. Samplings were made from control and toxicant chambers at 4hr interval. The experiment lasted for 96 hrs. Flow rate (1 lt/hr) was adjusted in such a way to replace the water in the respiratory chamber as well as the control chamber. Winkler's method (Strickland and Parsons, 1971) was adopted to measure the oxygen content. The respiratory rate was expressed in ml oxygen/g dry weight/hr.

6.3.3. RESULTS

The observed results are shown in Fig (24 A, B). The respiratory activity of both the species was found to decrease with an increase in the test concentration of endosulfan, whereas increased respiratory rate was observed in dimethoate exposed animals when compared with the control. In acute concentration of endosulfan (10 ppb *T. ala*; 6 ppb *T. cuspis* the rate of oxygen consumption in both species was significantly decreased by 50% over the control. In Tellinids exposed to 2 ppb of endosulfan the rate of oxygen consumption was reduced, followed by 4 ppb in *T. ala* and 3 ppb followed by 3 ppb in *T. cuspis.* When compared to endosulfan, in dimethoate exposed animals, at the acute (15 ppb; 10 ppb) and sublethal concentrations, oxygen consumption was found to be increased. The elevated respiratory rate in *T. ala* exposed to 8 ppb followed by 4 ppb in *T. cuspis* 5 ppb followed by 3 ppb was observed. In

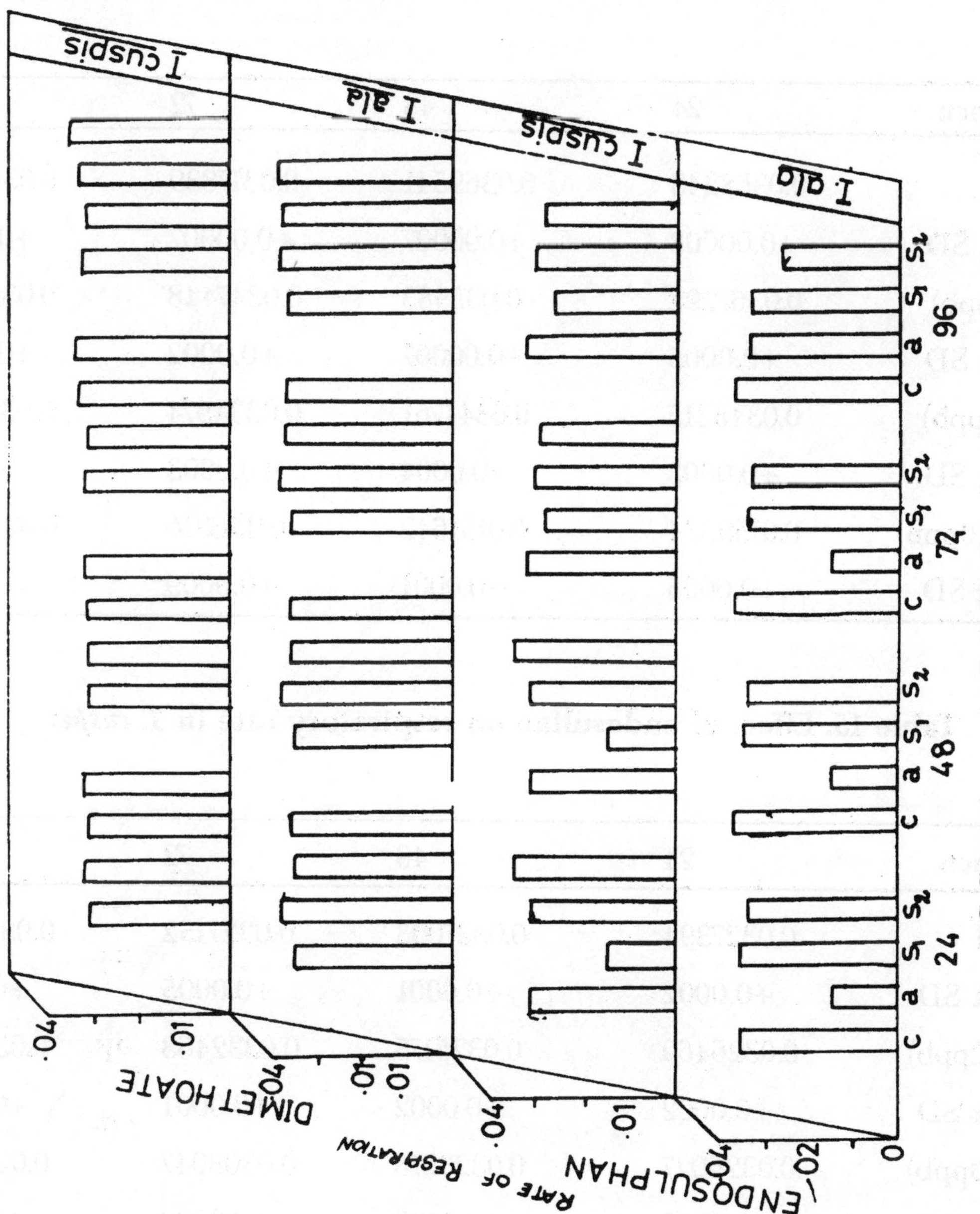

Fig. 24 A, B : Shows the rate of oxygen consumption by *T. ala / T. cuspis* exposed to different concentration of endosulfan and dimethoate.

Table 14. Effect of endosulf an on respiratory rate in *T. ala* (ml/O_2 /g dry wt/hr)

Test Concn.	24	48	72	96
Control	0.0368243	0.0369541	0.0317839	0.0376386
Mean & SD	+0.00008	+0.00007	+0.00007	+0.0005
Sub l(2ppb)	0.0367298	0.035983	0.0347448	0.0347699
Mean & SD	+0.0005	+0.00007	+0.0002	+0.0004
Sub 2(4ppb)	0.0348315	0.0344761	0.0334974	0.0332671
Mean & SD	+0.0003	+0.0004	+0.0003	+0.004
Acute (10ppb)	0.0159384	0.0151642	0.0156316	0.0265585
Mean & SD	+0.0005	+0.0001	+0.0009	+0.0001

Table 15. Effect of endosulfan on respiratory rate in *T. cuspis*

Test concn.	24	48	72	96
Control	0.0327394	0.0327003	0.0327152	0.0334249
Mean & SD	+0.0002	+0.0001	+0.0005	+0.0002
Sub 1 (2ppb)	0.0326469	0.0325175	0.0332463	0.0335317
Mean & SD	+0.0002	+0.0002	+0.0001	+0.0004
Sub 2 (3ppb)	0.0326027	0.0323278	0.0308047	0.0296814
Mean & SD	+0.0003	+0.0002	+0.0002	+0.0001
Acute (6 ppb)	0.0155965	0.0154169	0.0156316	0.0265585
Mean & SD	+0.0002	+0.0002	+0.0002	+0.0002

Table 16. Effect of Dimethoate on respiratory rate in *T. ala.*

Test Concn.	24	48	72	96
Control	0.0368243	0.0369541	0.0371783	0.0376386
Mean & SD	+0.00008	+0.00007	+0.00007	+0.0005
Sub l(4ppb)	0.036667	0.031662	0.0382939	0.0845829
Mean & SD	+0.0003	+0.0002	+0.0002	+0.0004
Sub 2(8ppb)	0.0376319	0.0376936	0.0386325	0.0394727
Mean & SD	+0.0005	+0.0002	+0.0004	+0.0003
Acute (15ppb)	0.0394869	0.0397079	0.0395893	0.0396646
Mean & SD	+0.0002	+0.0002	+0.0003	+0.0004

Table 17. Effect of Dimethoate on respiratory rate in *T. cuspis.*

Test Concn.	24	48	72	96
Control	0.0325845	0.0324621	0.0327221	0.0333238
Mean & SD	+0.0001	+0.0002	+0.0004	+0.0003
Sub l(3ppb)	0.0327897	0.0326454	0.0344649	0.0347521
Mean & SD	+0.0002	+0.0002	+0.0004	+0.0003
Sub 2(5ppb)	0.0334129	0.0336034	0.0354334	0.0363136
Mean & SD	+0.0002	+0.0004	+0.0002	+0.0002
Accute (10ppb)	0.0334426	0.0324481	0.03284	0.0324576
Mean & SD	+0.0002	+0.0003	+0.0001	+0.0004

Table 18. 't' values for the differences in respiratory rate of *T. ala* exposed to different concentrations of endosulfan.

24 hrs	2ppb	4ppb	10ppb
Control	0.2314	0.2475	4.5*
2ppb		0.2	2
4ppb			3*
48 hrs	2ppb	4ppb	10ppb
Control	0.17	0.2	5.4*
2ppb		0.15	2
4ppb			1.5
72 hrs	2ppb	4ppb	10ppb
Control	0.33	0.5	4.1*
2ppb		0.2	3*
4ppb			3 *
96 hrs	2ppb	4ppb	10ppb
Control	0.24	0.15	1.05
2ppb		0.06	1
4ppb			0.25

* indicates significant difference from control value ($P > 0.05$).

Table 19. 't' values for the difference in respiratory rate exposed to different concentrations of endosulfan.

24 hrs	2ppb	3ppb	6ppb
Control .	0.007	0.46	2.07
2ppb		0.004	3*
3ppb			1.5*
48 hrs	2ppb	3ppb	6ppb
Control	0.0074	0.5	2.5*
2ppb		0.03	3*
3ppb			3*
72 hrs	2ppb	3ppb	6ppb
Control	0.11	0.22	0.33
2ppb		0.4	0.6
3ppb			0. 2
96 hrs	2ppb	3ppb	6ppb
Control	0.12	0.5	0.71
2ppb		0.35	1
3ppb			0. 3

* indicates significant difference from control value ($P < 0.05$).

Table 20. 't' values for the differences in respiratory rate of *T. ala* exposed to different concentrations of Dimethoate.

24 hrs	4ppb	8ppb	15ppb
Control	0.02	0.02	0.33
4ppb		0.1	0.1
8ppb			0.15
48 hrs	4ppb	8ppb	15ppb
Control	0.02	0.02	0.33
4ppb		1	3*
8ppb			0.3
72 hrs	4ppb	8ppb	15ppb
Control	0.03	0.09	0.33
4ppb		0.03	0.1
8ppb			0.1
96 hrs	4ppb	8ppb	15ppb
Control	0.17	0.2	0.31
4ppb		0.1	0.1
8ppb			0.015

* indicates significant different from control value. ($P < 0.05$).

Table 21. 't' values for the differences in respiratory rate of *T. cuspis* exposed to different concentrations of Dimethoate.

24 hrs	3ppb	5ppb	10ppb
Control	0.006	0.083	0.083
3ppb		0.1	0.1
5ppb			0.1
48 hrs	3ppb	5ppb	10ppb
Control	0.05	0.12	0.04
3ppb		0.1	0.015
5ppb			0.1
72 hrs	3ppb	5ppb	10ppb
Control	0.1	0.12	0.06
3ppb		0.1	0.15
5ppb			0.4
96 hrs	3ppb	5ppb	10ppb
Control	0.18	0.15	0.02
3ppb		0.15	0.2
4ppb			0.35

dimethoate exposed Tellinids, the oxygen consumption was increased than the control at all the concentrations tested. Increase in respiratory rate was more in acute than sublethal exposure. In all the two sublethal concentrations and acute concentration tested, the oxygen consumption was found to increase upto 96hr in dimethoate exposed Tellinids and decrease in endosulfan exposed Tellinids.

Of the two pesticides tested against *T. cuspis* and *T. ala* respiratory rate was more affected by endosulfan than dimethoate. It is known fact that organochlorine is more toxic than organophosphorus substances. Further *T. ala* was found to be more susceptible to both the pesticides tested. The highest value was observed in 15 ppb of dimethoate (0.04 ml 02/g dry wt/hr) on exposure for 96 hr.. Lowest value was observed in 10 ppb of endosulfan 0.02 ml 02/g dry wt/hr on 96th hr of exposure. Though there are considerable changes in respiratory rate in the animals exposed to pesticides, they are not statistically significant in all the concentrations except in acute concentration (10 ppb) of endosulfan (Table 14 to 21).

6.3.4. DISCUSSION

In Tellinids exposed to pesticides normal metabolic activities seem to be affected. Any toxic treatment would adversely hamper its survival. While acute concentrations are lethal, chronic concentration leads to physiological damage which is as equal as mortality in the long run. The physiological damage includes general metabolism and respiration. Changes in oxygen consumption are valuable indicators of general environmental stress and are being used widely to evaluate the metabolic stress caused by change in environmental conditions brought by both natural and anthropogenic processes. The parameter commonly used to assess physiological adjustment of an organism to stress is oxygen consumption. In the present investigation changes in the oxygen consumption due to acute and sublethal levels of pesticides were studied.

The oxygen consumption decreased with a increase in the test concemtration of endosulfan. Similar decrease (MacInnes and Thurberg, 1973) and increase *(Thurberg et al* 1974) in oxygen consumption have been reported for various aquatic organisms after exposure to silver. Change in oxygen consumption of silver exposed bivalves was reported by Thurberg *et al* (1974) and of silver exposed gastropod mollusc *Nassa obsoletus* by Maclnnes and Thurberg (1973 a). He found that silver depressed whole animal oxygen consumption in this mud snail.

Thurberg *et al* (1974) reported the sublethal effects of silver on oxygen consumption of four bivalve, species held at various salinities. They observed that silver altered the respiration in the majority of cases, in each alteration an elevation in oxygen consumption was noted; this is similar to the results of the present investigation.

Whereas Scott and Major (1972) reported that copper depressed respiration in *Mytilus edulis*. Working with the same species Brown and Newell (1972) also noted copper induced depression in oxygen consumption while noting that zinc had no effect. Thurberg *et. al* (1974) opined that changes in the oxygen consumption may vary from species to species, nature of pollutant and also with respect to the life stages. Emmanuel (1985) reported significant increase in the respiratory rate of *Anadara rhombea* exposed to different concentration of WSF of diesel. Stekoll *et al* (1980) have reported similar rise in respiratory rates of *Macoma balthica* on exposure to low and higher concentration of Prudhoe Bay crude oil. Increased rate of oxygen consumption has been observed in *Mya arenaria* exposed to low concentration of PHC. (Fong, 1976; Gilfillan et al 1976, 77; Stainken, 1978). Approximately a two fold increase in respiratory rate with rising exposure concentration between 0 and 20% WSFs had been reported by Neff *et al* (1976) in shrimp.

In the present study, the respiratory activity of *T. ala* and *T. cuspis* sas found to decrease in higher concentration of endosulfan and increase in dimethoate. Similarly in contrast to the increase in respiratory rate by the organisms to PHCs, Dunning and Major (1974) and Gilfillan (1975) have reported a reduction in respiratory rate in *Mytilus edulis* exposed to higher concentration of PHC as a result of valve closure and poor ventilation.

Pollution increases oxygen consumption rate (Oksama, personal comm.,) or reduces oxygen consumption rate (Edwards, 1978) depending on the chemical nature and concentration of pollutant. Accumulating evidences indicate that organochlorines may interfere with respiratory enzymes (O'Brien, 1967).

Disruption of normal physiological processes by sublethal concentrations of pollutants may ultimately be as important as outright mortality. The perpetuation of a species depends on the ability of its individuals not only to survive but also to grow, reproduce and respond to changes in their environment. Many marine invertebrates are able to maintain their rate of oxygen consumption in water of declining oxygen content by either increasing the ventilatory rate and or by increasing the efficiency of oxygen uptake (Newell, 1979).

Pollutants like heavy metals and pesticides are known to affect marine animals in a variety of ways. They disrupt enzyme systems (Jackim *et al* 1970; Gould and Karolus, 1974) cause tissue and organ damage resulting in abnormal function and provoke behavioural changes (Gardner and Yevich, 1970; Newman and MacLean, 1974). Maclnnes and Thurberg *et al.* (1975) reported that the increased metabolic activity of surf clams exposed to silver demonstrates physiological disruption and an abnormal expenditure of energy. When naturally occurring stress factors threaten the internal milieu in marine organisms it is stabilized by regulatory processes resulting in an increased metabolic demand. This raised oxygen demand is quickly met by increased external respiration (Hagerman, 1970,76). Stekoll *et al* (1980) have also concluded that the basal metabolic rate might be altered due to the lack of food in *Macoma balthica*. And this alteration in metabolic rate may increase the respiratory rate as the organisms were not fed prior to the exposure.

In the present study in each concentration of dimethoate the oxygen consumption increased upto 96hr and it might be due to increased metabolic rate to overcome the stress caused by pesticides. The fall in oxygen consumption at 96hr observed in endosulfan exposed Tellinids might be due to interference with respiratory enzymes. It was also observed that *T. ala* was found to be more susceptible than T. *cuspis*. Stirling (1970) reported that larger individuals of Tellinids were more sensitive to toxins than small individuals. Similar observation was also obtained in the present study.

6.4. EFFECT OF PESTICIDES (ENDOSULFAN AND DIMETHOATE) ON FILTRATION RATE OF BIVALVES Tellina ala AND Tellina cuspis (Hanley) (MOLLUSCA : BIVALVIA)

6.4.1. INTRODUCTION

Adult bivalves are either sessile, sedentary or burrowing forms. In many bivalves, their life functions are carried out in an all - or - none fashion contingent upon whether or not the animal is pumping seawater through its mantle cavity (ventilating). The behaviour, physiology, growth and reproduction of bivalves are much interdependent with respect to ventilation (Sunil Kumar Mohamed, 1980). The filtration rate of bivalves, that is the volume of water drawn into the animal's body is known to be influenced by environmental parameters such as salinity, temperature, dissolved oxygen and concentration of suspended matter (Cole and Hepper, 1954; Badman, 1975; Mane, 1975).

The filtration rate of suspension feeding animals can be estimated by measuring the rate at which the animals clear a suspension of particulate material. The ability of bivalves to absorb substances from the surrounding water and concentrate them in their tissues enables them to be treated as pollution indicators (Phillips, 1976; Zaroogian, 1980).

It is well known that filtration in bivalves serves several purposes including respiration, feeding and discharge of gametes and wastes. A pollutant which affects filtration rate may be expected to exert multiple effects on organisms. Reports are available on the effects of various pollutants on physiological activities like feeding, respiration, excretion and various other life activities such as ciliary activity, burrowing and byssal activity in different organisms. Davis (1961) demonstrated both positive and negative effects of pesticides on the growth and survival of juveniles of *Crassostrea virginica* and *Venus mercenaria*. Loosanoff (1947) reported effects of DDT on the settlement, growth and survival of oysters. Effect *of* pesticides on pumping rate, shell movements and growth in *C. virginica* was reported by Butler *et al.* (1962). Roberts (1975) studied the effect of endosulfan, DDT and dieldrin on respiration, heart beat and burrowing of some marine bivalves. Kittredge *et al.* (1974) studied the effect of naphthalene on ciliary activity. However, information on the effects of critical pollutants on filtration rate is limited.

Table 22.

Species	Pesticide concentration							
	Endosulfan				Dimethoate			
	High	SL_1	SL_2	SL_3	Acute	SL_1	SL_2	SL_3
Tellina ala	10 ppb	4ppb	2ppb	1ppb	15ppb	8ppb	4ppb	2ppb
Tellina cuspis	6ppb	3ppb	2ppb	Ippb	10ppb	5ppb	3ppb	1ppb

The technique applied to measure the rate of filtration needs very little apparatus and seems to have wide application because neutral red in estuarine water is absorbed and retained by the gills of the lamellibranchs. The method followed in the present study was the same as used by Cole and Hepper (1954) and Nagabhushanam (1956) using neutral red solution. The filtration rate was estimated by measuring the rate of absorption of neutral red (Abel, 1976 and Badman, 1975). Though much data are available on the toxic effects of heavy metals to many species on physiological aspects, information on pesticidal effect on rate of filtration in bivalves is scanty. Hence an attempt was made to evaluate the effect of pesticides on the rate of filtration in two Tellinids.

6.4.2. MATERIALS AND METHODS

The collection, acclimation and biological condition of the test animal, physical, chemical characteristics of the test water, preparation of different concentrations of pesticides and other experimental conditions are described earlier (6.2). Adults of uniform size with extended siphons and valves that opened sufficiently to display the edge of the mantle were selected.

T. ala and *T. cuspis* were exposed to acute concentration of 10 ppb and 6 ppb and sublethal concentrations of 4 ppb, 2 ppb and 1 ppb (*T. ala*), 3 ppb, 2 ppb and 1 ppb (*T. cuspis)* of endosulfan and dimethoate for a period of one hour and controls for the both species were also maintained for each set of experiment. Glass beakers (500 ml) were arranged on a table with white background. This would enable a clear vision to study the siphoning behaviour of the animal. Forceps was used for handling the animal so as to minimize the stress for the animal. No feed was given during the experimental period. 6 specimens each of *T. ala* and *T. cuspis* were placed separately in beakers containing 400 ml of toxic medium, whereas the control animals were placed in the pesticide free estuarine water which received acetone only. The dye solution was prepared by dissolving 1 mg of neutral red in 100 ml distilled water, diluting the resulting solution to high strength sea water to give a solution of 20%o salinity with sufficient dye to give a concentration of approximately 1 ml/1. 4 ml of 1% neutral red was added to each test container and gently mixed by stirring. 10 ml of water sample was immediately pipetted out from each test container. Then subsequent samplings were made at a regular interval of 15 mts for an hour, and after making sure that the animals were filtering actively. After adding a few drops of dilute HC1 (50%) to the samples, the dye concentration in the samples was determined spectrophotometrically at 425 nm in a Hitachi double beam spectrophotometer (Model 2203). The readings were expressed in percent concentration. In each concentration 6 animals were used.

6.4.3. RESULTS

The variations in the rate of filtration with varying concentrations of the pesticides are presented in (Fig. 25 A). Control organisms filtered neutral red, suspended in the medium actively, whereas the rate of filtration was found to be altered in animals exposed to pesticides (Fig. 25 B; Pl. 25 A, B). Both the species of Tellinids exposed to high concentrations of pesticides, exhibited a reduced activity and a very low pumping rate and in low concentration increased activity was observed, when compared to the control. The present result revealed that in the organisms exposed to 10 ppb and 6ppb of endosulfan and (15 ppb and 10 ppb of dimethoate, the efficiency of the filtering capacity was significantly ($P > 0.05$) reduced from 29% to 2.5% (*T. cuspis*) 27% to 2.2% (*T. ala*); 31% to 3% (*T. ala*) and 29% to 2.8% in *T. cuspis* respectively (Table 23, 24). At the high sublethal concentrations both pesticides exposed Tellinids (*T. ala* and T. *cuspis)* the filtration rate was slightly increased than the acute and reached half the level of control value (Fig. 25 A). In *T. ala* exposed to (4 ppb) of endosulfan the rate of filtration was reduced to 12%; and in *T. cuspis* (3 ppb) the pumping rate was reduced to 10% whereas in dimethoate exposed Tellinids, the filtration rate was reduced to 16% and 13% in *T. ala* (8 ppb) and *T. cuspis* (5 ppb) respectively.

At median sublethal concentrations of both pesticides the rate of filtration was higher than acute and high sublethal levels among exposed Tellinids. The rate of filtration in dimethoate exposed *T. ala* is 18% (4 ppb) and of T. *cuspis* (3 ppb) *is* 16%. Whereas it is increased as 14% in *T. ala* and 12% in *T. cuspis* exposed to sublethal concentrations of endosulfan (*T. ala* 2 ppb, *T. cuspis* 2 ppb) after exposure of 1 hr (Table 25, 26).

At low sublethal concentrations of endosulfan and dimethoate the animals were active and the percentage of rate of filtration increased than controls. The percentage increased to 29.5% (1 ppb *T. ala*) and 31% (1 ppb *T. cuspis* in endosulfan exposure and it was observed as 31.5% and 34.4% in *T. ala* (2 ppb) and T. *cuspis* (1 ppb) respectively in dimethoate exposure.

The rate of filtration decreased exponentially with increasing concentrations of pesticides. The decreased was found to be statistically significant ($P > 0.05$) (Table 27, 28). Endosulfan was reported to be considerably more toxic than dimethoate which is confirmed from the present results. And generally the rate of filtration was found to be increased upto 30 mts. afterwards decline in pumping rate was observed at the end of one hour. *T. ala* was found to be affected more than *T. cuspis.* (29, 30)

6.4.4. DISCUSSION

Pesticides cause stress on the organism and disturb its physiological activities. Many investigations have been carried out on toxic effects on filtration rate of various lamellibranchs; Davis (1961) in *Crassostrea virginica;* Engel *et al.* (1972) in *Mercenaria merceneria;* Abraham *et* al. (1986) in *Villorita cyprinoides* and Nagarajan (1987) in *Meretrix casta.* Newell (1973) reported that the pollutant might cause physiological stress in exposed organisms as do the altered environmental parameters. Lund (1967) investigated the effects of petroleum fractions on the pumping and clearance rates of oysters (*C. virginica)* and found that filtration rate was the more sensitive physiological and behavioural index. Threshold concentration of 3-6% of petroleum fraction affected the rate of filtration in oysters. The filtration rates of the soft shelled clam *Mya arenaria* increased at the low concentration (0.9%) of WSF of crude oil (Anderson, 1972). Similar observation was noted in the present study also. At low concentrations of pesticide (1 ppb) (*T. ala* endosulfan), (1 ppb) (*T. cuspis* endosulfan) (2 ppb *T. ala* dimethoate and 1 ppb *T. cuspis* dimethoate) the rate of filtration was slightly increased than that observed

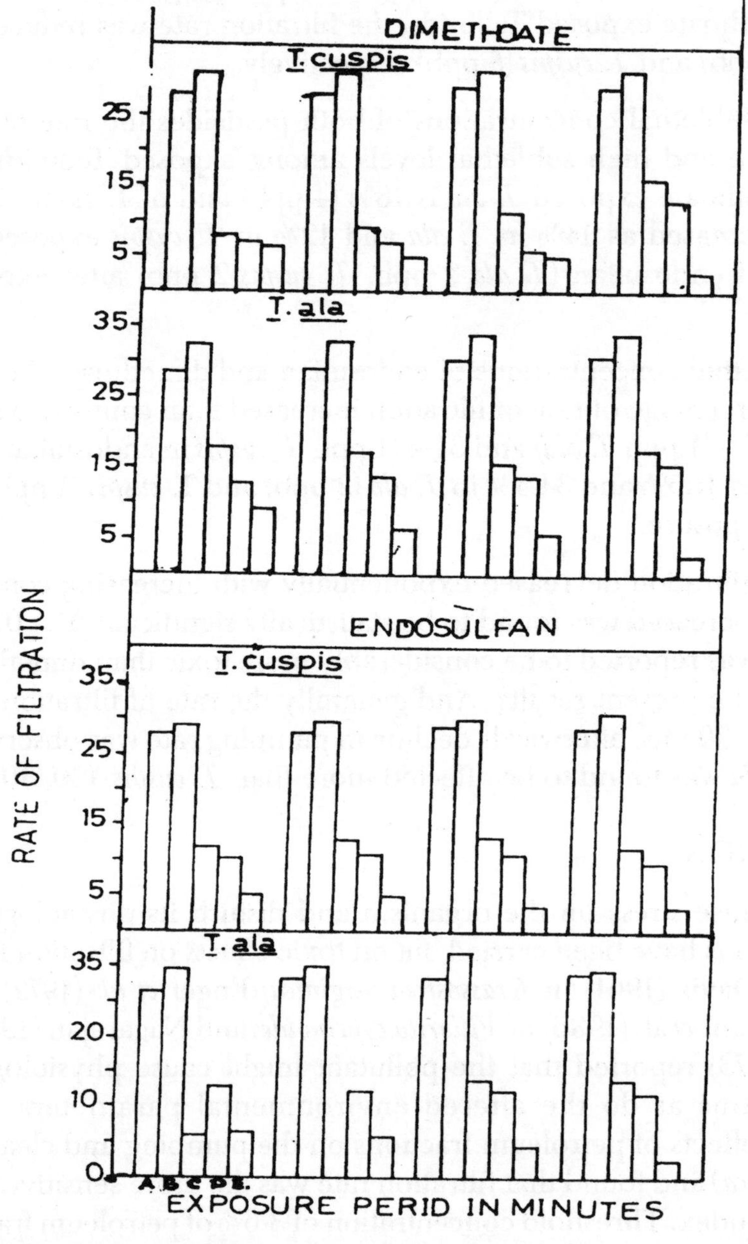

Fig. 25 A : *T. ala.* Rate of filtration at different concentrations of endosulfan and dimethoate showing linear relationship.

Fig. 25 B : *T. cuspis* Rate of filtration at different concentrations of endosulfan and dimethoate showing linear relationship.

A. Control B. Low SL; C. Medium SL; O. High SL; E. Acute.

Table 23. Effect of endosulfan on (Rate of filtration) in *T. ala*.

T. ala		15	30	45	60
Control	Mean	26.38	28.08	28.23	28.41
	SD	+0.38	+0.23	+0.47	+0.5
Low SL (1ppb)	Mean	28.96	29.33	31.25	30.03
	SD	+0.58	+0.68	+0.85	+1.39
Median SL (2ppb)	Mean	12.436	11.5	13.46	13.87
	SD	+0.4	+0.3	+0.4	+0.17
High SL (4ppb)	Mean	6	12.40	12.33	12.20
	SD	+0.3	+0.3	+0.3	+0.2
Acute (10ppb)	Mean	6.15	5.95	4.3	2.25
	SD	+0.2	+0.3	+0.2	+0.2

Table 24. Effect of endosulfan on Rate of filtration

T. cuspis		15	30	45	60
Control	Mean	27.58	27.85	28.65	28.7
	SD	+0.3	+0.23	+0.42	+0.2
Low SL (1 ppb)	Mean	11.8	12.8	13.33	11.75
	SD	+0.2	+0.2	+0.2	+0.3
Median SL (2 ppb)	Mean	10.26	28.72	29.73	30.87
	SD	+0.2	+0.4	+0.2	+0.2
High SL (3 ppb)	Mean	28.75	10.58	10.53	9.52
	SD	+0.2	+0.24	+0.2	+0.3
Acute (6 ppb)	Mean	5	4.46	3.32	2.47
	SD	+0.13	+0.31	+0.21	+0.31

Table 25. Effect of dimethoate on (Rate of filtration)

T. ala		15 mts	30 mts	45mts	60mts
Control	Mean	28.28	28.95	30.5	30.76
	SD	+0.24	+0.17	+0.33	+0.4
Low SL (2ppb)	Mean	32.22	32.57	33.68	34.45
	SD	+0.2	+0.4	+0.33	+0.35
Median SL (4ppb)	Mean	17.73	17.13	15.71	17.73
	SD	+0.4	+0.36	+0.37	+0.37
High SL (8ppb)	Mean	12.46	13.5	13.91	14.98
	SD	+0.3	+0.38	+0.2	+0.64
Acute (15 ppb)	Mean	8.92	5.95	5.7	2.76
	SD	+0.22	+0.3	+0.45	+0.23

Table 26. Effect of dimethoate on (Rate of filtration)

T. cuspis		15 mts	30 mts	45mts	60 mts
Control	Mean	27.6	27.49	28.45	28.56
	SD	+0.25	+0.64	+0.47	+0.29
Low (1 ppb)	Mean	30.65	30.73	30.88	30.93
SL	SD	+0.22	+0.37	+0.25	+0.44
Median (3ppb)	Mean	7.35	8.65	10.86	15.66
SL	SD	+0.29	+0.38	+0.31	+0.37
High (5ppb)	Mean	6.91	6.63	5.4	12.46
SL	SD	+0.22	+0.13	+0.22	+0.39
Acute (10 ppb)	Mean	6.23	4.98	4.9	2.73
	SD	+0.22	+0.11	+0.13	+0.14

Table 27. 't' values for the differences in filtration rate of *T. ala.* exposed to different concentrations of endosulfan.

15 mts	1 ppb	2 ppb	4 ppb	10 ppb
Control	6.4	60.5	41.43	64.64
1ppb		60.24	40.9	63.73
				*
2ppb			18.86	0.52
4ppb				1.9.76
30 mts	1 ppb	2 ppb	4 ppb	10 ppb
Control	3.21	55.78	52.75	74.45
1ppb		44.11	41.88	57.84
2ppb			2.81	17.47
4ppb				20.31
45 mts	1 ppb	2 ppb	4 ppb	10 ppb
Control	6.44	38.78	44.37	71.60
1ppb		39	43.12	64.82
2ppb			2.586	29.38
4ppb				27.82
60 mts	1 ppb	2 ppb	4 ppb	10 ppb
Control	58.71	4.85	66.44	89.97
Ippb		31.81	32.85	53.45
2ppb			5.45	46.81
4ppb				31.90

* indicates insignificant difference from control value ($P > 0.05$).

Table 28. 't' values for the differences in filtration rate of *T. cuspis* exposed to different concentration of endosulfan.

	Low	Median	High	Acute
15 mts	1	2	3	6
Control	54.66	4.05	67.66	84.35
1ppb		65.22	5.93	28.70
2ppb			71.16	100.32
3ppb				22.22
	Low	Median	High	Acute
30 mts	1	2	3	6
Control	52.13	2.54	8.196	73.36
1ppb		50.13	8.19	2.88
2ppb			55.13	70.03
3ppb				20.36
	Low	Median	High	Acute
45 mts	1	2	3	6
Control	59.33	4.18	70.16	21.40
1ppb		63.11	10.78	38.53
2ppb			73.89	101.64
3ppb				27.76
	Low	Median	High	Acute
60mts	1	2	3	6
Control	58.71	4.85	66.44	89.97
1ppb		66.24	7.02	32.14
2ppb			73.96	98.382
3ppb				24.42

All values show significant difference from control value ($P > 0.05$).

Table 29. 't' values for the differences in filtration rate of *T, ala* exposed to different concentrations of Dimethoate.

15mts	2	4	8	15
C	14.55	32.31	52.73	69.93
2ppb		45.82	68.41	88.06
4ppb			15.43	27.41
8ppb				12.02
30mts	2	4	8	15
C	11.74	39.77	51.01	82.18
2ppb		43.38	52.90	77.93
4ppb			10.34	33.70
8ppb				12.95
45mts	2	4	8	15
C	9.58	43.29	55.8	68.79
2ppb		52.61	64.14	77.59
4ppb			4.87	27.06
8ppb				24.95
60mts	2	4	8	15
C	10.44	36.36	37.90	86.41
2ppb		48.26	47.92	101.92
4ppb			6.70	47.34
8ppb				32.09

All values indicate significant difference from control value. (P > 0.05}.

Table 30. 't' values for the differences in filtration rate of *T. cuspis* exposed to different concentrations of Dimethoate.

15mts	1ppb	3ppb	5ppb	10ppb
Control	10.90	67.5	73.92	76.34
1ppb		79.93	87.67	53.20
3ppb			1.50*	2.33*
5ppb				2.51
30mts	1ppb	3ppb	5ppb	10ppb
Control	7.89	54.47	58.23	63.66
1ppb		87.74	83.48	91.04
3ppb			6.93	12.84
5ppb				8.25
45mts	1ppb	3ppb	5ppb	10ppb
Control	7.01	48.79	67.97	74.46
1ppb		65.52	91.03	90.70
3ppb			8.99	44.35
5ppb				32.74
60mts	1ppb	3ppb	5ppb	10ppb
Control	6.79	38.89	32.74	96.48
1ppb		41.98	49.67	90.70
lppb			8.99	44.34
5ppb				32.72

* indicate insignificant difference from control value. (P > 0.05).

in the control. This could be due to excitement. It was also observed that higher concentration of pesticides arrested physiological activities. The rate of filtration decreased with an increase in the test concentration of pesticides. This may be due to atrophy of gill and labial pulp tissues affected by pesticides. Similar observations were reported in *Mytilus edulis* exposed to copper and zinc (Brown and Newell, 1972); *Meretrix casta* exposed to endosulfan (Nagarajan, 1987) and in *Villorita cyprinoides* exposed to heavy metals (Abraham *et al.* 1986).

Miller and Kinter (1977) found that the organochlorines such as DDT, disrupt the membrane transport function in sensitive species. Neutral red is reported to have the effect of increasing valve activity of mussels when saturated with the stain due to prolonged exposure (Salanki and Lukacsovics, 1967). Therefore, it was felt desirable to restrict the observations to the first 2-3 hr. The rate of filtration of the animal was not uniform.

The results of the present study clearly indicate that the pesticides endosulfan and dimethoate have a potential effect on the physiological activity (rate of filtration) of Tellinids. Endosulfan was observed to be considerably more toxic than dimethoate. Stirling (1970) reported that larger individuals of *Tellina tenuis* were found to be more sensitive to toxins than smaller individuals which was vice versa in *M. edulis* In the present study also *T. ala* which is larger in size than *T. cuspis* showed more sensitivity to pesticides.

7

HISTOPATHOLOGICAL EFFECTS OF ENDO-SULFAN ON DIFFERENT BODY TISSUES OF THE *Tellina ala*

7.1. INTRODUCTION

There is growing body of literature on the biological and toxicological effects of pollutants and relative toxicities of contaminants to several species of marine organisms and number of investigations have been carried out on the effect of various pollutants on behavioural, biochemical and physiological changes of various organisms. However, informations on the effects of pesticides on histopathological changes are limited. The study of histopathology effected by pollutants is an important basic effort leading to our understanding of true impact of pollutants on the ecosystem. The secondary effects of pollutants and the extent of damage caused to vital tissues can be studied by histopathological observations. The aim of the histopathological studies was mainly to assess the extent of internal pathological lesions caused due to accumulation of pesticides.

The present chapter deals with the hitopathological effects of endosulfan (acute 10 ppb and sublethal) on different body tissues (siphon, foot, oocyte, digestive diverticula and gills) of *Tellina ala*.

7.2. MATERIALS AND METHODS

The collection, acclimatisation, biological condition of the test organisms and the physical, chemical characteristics of the test water used in the present study have been described in sec (6.2).

Tellina ala were exposed to acute concentration (10 ppb) and sublethal concentration (2 ppb) of endosulfan. After 96 hr of acute exposure the different soft tissues like siphon, foot, oocyte, digestive diverticula and the gill of experimental and control animals were dissected out under physiological saline solution and immediately transferred to Bouin's fixative. The permanent preparations of the slides were made following the procedure of paraffin method, serial sections of 6-8 u thickness were deparaffinised and stained in Heidenhain' s haematoxylin using eosin as counterstain.

7.3. RESULTS

In the present study, attention was paid to gill, foot, siphon, oocyte and digestive diverticula to see whether endosulfan had any effect and also to know if there is any correlation between the amount of concentration of endosulfan and level of damage of cells.

Control : Gill

The paired gills or ctenidia are suspended in the mantle cavity on either side of the visceral mass. The ascending limb of the gill lamella curved at the tip to form the blood grooves. The frontal area is covered by a wide row of frontal cilia. Epithelium of the frontal region shows the presence of darkly staining mucous cells. The membranous basal portion contains longitudinal and transverse muscle fibers. As in other eulamellibranchs, the ctenidia are composed of the filaments, interfilamentar junctions, ostia and water tubers. Internally the filaments are supported by rods and the epithelium consists of a single layer of ciliated cells, the ciliation being restricted to the lateral, laterofrontal and frontal regions (Pl. 26 A, B, 27 A, 27 D)

Acute (10 ppb)

In acute exposure (Pl. 27 B) severe necrosis in lamellar region was noticed. The mucous *cells* and ciliated cells were affected. The blood groove and blood space were extremely destroyed. Changes in haemocytes, destruction of epithelial layer and separation of lamella were observed. In acute dosage entire shape was completely lost due to severe necrosis (Pl. 27 C).

Sublethal exposure (2 ppb)

In sublethal exposure (Pl. 27 E) the overall histopathological changes are less when compared to acute exposure cells. The shape was not much affected. Invasion of gill wall was observed.

Siphon : Control

Two siphons (an inhalent and exhalent one) occur as muscular tubes both formed by the fusion of the mantle folds posteriorly. The siphon of the animals maintained as control shows an inner and outer surfaces of the siphonal walls; they are covered by a layer of cubical epithelium. Longitudinal and circular layer of muscles occur internally in between the epithelium of the siphonal walls (Pl. 28 A, 29 A, 29 D).

***Accute* (10 ppb)**

Severe damages were observed in the siphon due to acute exposure of endosulfan (Pl. 28 B, 29 B, 29 E). The outer and inner siphonal walls became irregular and shrinked. The cells of the cubical epithelium became elevated (Fig. 11). Degeneration of outer cuticle was observed. The circular layer of muscle disappeared and fusion of circular muscles were also observed.

***Sublethal* (2 ppb)**

Whereas in the sublethal exposure the damages are less when compared to acute exposure; only the circular muscle was found degenerated (Pl. 28 C, 29 C).

The cross section of treated siphon showed dilation of outer cuticle layer (Pl. 5, 6). Degeneration of cubical cells, rupture of cell wall and disappearance of nuclei were also observed (Pl. 29 F).

Control : Foot

The epithelium of foot consists of cubical cells and beneath the epithelium are seen several layers of foot muscles (Pl. 30 A, 30 D). The foot is used very actively as in some quick burrowing clams like *Donax* sp. and other tellinids (Yonge, 1949.). The foot is highly extendable and is provided with paired anterior and posterior pedal retractors.

***Acute* (10 ppb) and *Sublethal exposure* (2 ppb)**

When exposed to the acute endosulfan concentration, the foot showed the following histopathological changes, degeneration of foot muscles and other epithelial cells (Pl. 30 B, 30 C, 30 E, 30 F) the cross section showed irregular and elongated shape of foot. Vacuolization and degeneration of foot muscles was only observed in the sublethal exposure (Pl. 30 G, 30 H).

Gonad (Oocytes) Control

The gonad consists of numerous round or oval follicles which increase in size during maturation, occupying the connective tissue space between the outer wall and viscera. Each follicle consisting of an outer sheath and a follicle wall, houses the developing green cells. Each follicle is surrounded by an outer follicular sheath (Pl. 31 A).

Acute (10 ppb)

In the acute exposure the outer follicular sheath disintegrate and free nuclei were observed. Gonadial tumors were formed. The follicles were jammed, causing distension of *the* follicles. The nuclei became irregular in shape. The normal architecture of the oocytes were destroyed (Pl. 17). The histopathological observations showed the severity of effects of the pesticide was more marked at LC_{50} than sublethal exposure. This might be due to the penetration of pesticide deep into the follicles of gonad tissue (Pl. 31 B, C, D, E, F)

Digestive diverticula (Control)

Digestive diverticula consists of numerous blindly ending tubules which open into the stomach by a system of primary and secondary ducts. The tubules of digestive diverticula are rounded in cross section (Pl. 32 A). The cells are cubical with very thin wall. Each tubule consists of largely staining vacuolated secretary and absorptive cells (Pl. 32 C, D).

In acute exposure (Pl. 32 B), interlobular connective tissue lost its original shape and thereby affected the digestive cells. Endosulfan affected the vacuolated cells and secretary cells which lost their shape and became irregular. Degeneration of cell wall and free nuclei were observed.

Whereas in the sublethal exposure the nuclei disintegrate and cell walls are present. Shrinkage of small darkly stained generative cells were observed. Size of the lumen is comparatively small than that observed to the acute exposure. The overall damage is comparatively less in sublethal exposure.

7.4. DISCUSSION

Only very few works are available on histopathological effects of pollutants on bivalves. In the present study the effect of endosulfan on the histological structures of different body parts is obvious. Endosulfan considerably damaged the structure of siphon, foot, gonad, digestive diverticula and gills. The results observed in the present study is similar to that of *Indonaia caeruleus* when exposed to endrin, DDT and malathion (Mane *et al.*, 1979) and of *Meretrix casta* exposed to endosulfan (Nagarajan, 1987).

Oysters exposed to a mixture of DDT, toxaphene and parathion (1 ppb of each) for long periods (duration) were examined and reported by Lowe *et al.* (1971). They reported that gonadal maturation was retarded in female oysters subjected to this mixture for 36 weeks. They also observed hyperplasia of the germinal epithelium. In the present study disintegration of germinal epithelium was noted. And also loss of its original shape was observed.

The epithelial cells of the digestive tubules were reduced in height in treated oysters (Lowe *et al.*, 1971). Loss of shape of digestive cells, degeneration of cell wall and free nuclei were observed after treatment with endosulfan. Similar observations were made in the present study also. In acute exposure of endosulfan, the interlobular connective tissue lost its original shape and thereby affected the digestive diverticula cells in *T. ala.* Degeneration of cell wall, reduction in the size of lumen of cells and free nuclei were also observed. Vijayamadhwan and Iwai (1975) reported in oysters that pollutants cause neoplastic cell invaded through the sinuses and blood vessels to other organs, where it developed and replaced the normal architecture of these organs. Nagarajan (1987) observed endosulfan to destroy the original shape of hepatopancreas and thereby affected the digestive cells.

Vacuolation, necrosis and cellwall disintegration were commonly encountered at all the exposures of endosulfan in fishes were observed by Eller (1971); Mathur and Rane (1979); Ramalingam and Srinivasa Reddy (1981) and Ratnaker and Aswathy (1979). It could be understood from the present study that histopathological results observed were very similar with the results observed in fishes. Pathological lesions induced by Zn and Cu in gill epithelium of *Ambassis commersoni* (Pragatheeswaran, 1987) and by endosulfan and heavy metals in *Oreochromis mossambicus* (Jambulingam, 1988). And by endosulfan in *Meretrix casta* (Nagarajan, 1987) were reported. They observed degeneration of gill epithelium, necrosis, edema of gill lamellae and destruction of blood grooves and blood spaces, which is very similar to the present results in *Tellina ala* exposed to endosulfan. Histopathological changes induced by cythion-malathion in the gonads of lamellibranch molluscs were reported by Muley and Mane (1987). Neff *et al.* (1987) observed histopathological effects of spilled oil in Arctic marine bivalve molluscs. They observed increased incidence of gill and digestive tract necrosis in *Mya* sp. Nagarajan (1987) reported that *Meretrix casta* endosulfan considerably affected the organs like gill, labial palp, intestine and hepatopancreas.

In the present study gills were found to be affected more than other parts. Their physiologic function was probably impaired by the presence of neoplastic cells. Siphon and gills seem to be affected more when compared to other parts. This is because their surfaces were completely exposed to endosulfan contaminated water. Generally there is a linear relationship between the concentration of toxic substances and cell damages. In all body tissues studied the level of damages is more in acute than in sublethal exposure. These histopathological changes will definitely lead into impairment in the physiological activities like respiration, filtration and reproduction, etc.

8

SUMMARY

Present investigation was undertaken to study the ecology breeding biology and biochemical composition of the Tellinids, *T. ala* and *T. cuspis.* In addition to the biological studies, the effect of two pesticides endosulfan and dimethoate on the physiology and histology of Tellinids were also studied.

Data on the environmental parameters in the study area were collected for the period June 1987- May 1988. Generally, rainfall is high during monsoon period (Oct.-Dec.) and low in other seasons. Monthly variations in temperature of atmosphere, surface water and sediment showed not much variations; however, the atmospheric temperature was higher than surface water and sediment temperature. Highest temperature was recorded during summer months and lowest during monsoon months. The pH fluctuated from 7.5 to 8. Salinity showed an inverse relationship with dissolved oxygen. The range of dissolved oxygen fluctuated between 4.5 ml/1 and 7 ml/1. Generally, the percentage of sand composition was high when compared to silt and clay. Sandy nature of sediment was observed in all months except in monsoon months. Population abundance of Tellinids were found to be high during summer and postmonsoon periods and low during monsoon period. The animal collection made showed that the population of *T. ala* was dominant than *T. cuspis.* In the present experiment study, *T. cuspis* showed normal activity in the salinity range of 23-26 ppt and in *T. ala* the range was 27 to 30 ppt. In the present study area *T. ala* has a distributional limit which is higher than *T. cuspis* and they were found to have a tolerance to salinity, fluctuations. Temperature is known to influence the critical salinity tolerance. At low salinity thermal tolerance was reduced in *T. ala* and in *T. cuspis.* Temperature tolerance experiment revealed that 28-32°C was optimum temperature for both species of Tellina.

The shell is thin, fragile and transversely elongated. It is equivalved and pale white in colour in *T. ala*. In *T. cuspis* the shell is deep pink in colour. *T. cuspis* is smaller than *T. a1a*. The shell shape is highly adapted for burrowing movements. The mantle lobes are paired, triangular structures. The demibranchs are quite large and the labial palps are small. The inner demibranch is very narrow and longer than outer demibranch. The ctenidia composed of filaments, interfilamentar junctions, interlamellar junctions, ostia and water tubes. The siphons are long, separate and mobile. Longitudinal and circular layer of muscles occur internally between the epithelium of the siphonal walls. The foot is relatively longer in *T. ala* than *T. cuspis*. It is somewhat elongated and axe shaped in *T. ala* while in *T. cuspis,* the foot is slightly blunt. The foot is highly extendable and is provided with paired anterior and posterior pedal retractors. The epithelium of the foot consists of cubical cells and beneath the epithelium are seen several layers of foot muscles. The most conspicuous features of the digestive system in these Tellinids are the very capacious stomach while the oesophagus which is relatively short and opens into the anterior part of the large stomach.

The digestive diverticula form a dense mass of brown tubules below the stomach and the ducts of digestive diverticula open into the food sorting caecum situated in the anterior region of the stomach. The combined crystalline style sac and midgut arise from the posteroventral portion of the stomach. The intestinal wall is composed of ciliated epithelial cells and less numbers of mucous cells. The rectum is thinner than the rest of the intestine. The gonads are well developed encircling the intestine and imbricated in the digestive gland.

The present study reveals that Tellinids have prolonged breeding season from February to September. *T. ala* attains first sexual maturity at 11-13 mm and 8-10 mm in *T. cuspis*. Chi-square analysis showed sex ratio 1:0.91 in *T. ala* and 1:0.84 in *T. cuspis* showing slight dominance of females in both the species. It was also observed that salinity seems to play an important role and the reproductive cycle starts from February onwards and no breeding of Tellinids was observed during monsoon period.

Results obtained from the biochemical studies on Tellinids showed the percentage of biochemical constituents were maximum during maturation period and minimum during spawning period. Increase in biochemical constituents during summer is associated with the proliferation of gonadial tissue in the visceral mass, and decrease in September mainly due to active spawning. And the percentage of biochemical constituents are higher in females than in males. The biochemical constituents are high in gonad and digestive diverticula. Least fluctuations in the biochemical constituents of mantle and adductor muscles were observed, The protein is the main constituent followed by lipid and carbohydrate. The percentage of nutrients in *T. ala* was greater than *T. cuspis*. It was observed in the present study that seasonal variations in biochemical constituents in Tellinids were mainly influenced by reproductive cycle and secondarily by environmental changes and availability of food.

The present investigation also reports toxicity of endosulfan (organochlorine) and dimethoate (organo-phosphate) on physiology (Rate of filtration and respiration) and histology of *T. ala* and *T. cuspis*. Endosulfan was found to be more toxic than dimethoate *T. ala* was found to be more sensitive to pesticide than *T. cuspis*. The rate of filtration decreased with an increase in the test concentration of pesticides.

The respiratory activity of both the species was found to decrease with an increase in the test concentration of endosulfan, whereas increased respiratory rate was observed in dimethoate exposed animals. Of the two pesticides tested against *T. cuspis* and *T. ala* respiratory rate was more affected by endosulfan than dimethoate. Further *T. ala* was found to be more susceptible to both pesticides tested.

Histopathological effects of endosulfan on different body tissues of *T. ala* showed considerable damage in the structure of siphon, foot, gonad (oocyte), digestive diverticula and gills. Histopathological changes like degeneration of outer epithelial layer, mucous cells and ciliated cells, destruction of blood groove and blood spaces, disappearance of cilia, severe necrosis of lamellar region and separation of gill lamella were observed in the gills of *T. ala* exposed to acute concentration of endosulfan. In sublethal concentration the shape of the gill was not much affected. Due to pesticide treatment in siphons, the outer and inner siphonal walls become shrinked and irregular in shape. Besides elevation of cubical layer of muscles in between the siphonal walls dilation of siphonal wall was also observed. In sublethal exposure the damages are less when compared to acute exposure; only the circular muscle was found degenerated. Pesticide exposed foot showed degeneration of foot muscle and outer epithelial layer. Endosulfan affected the vacuolated and secretory cells of digestive diverticula, which lost their shape and become irregular. In the present study siphon and gills were found to be affected more than other parts, since their surfaces were completely exposed to endosulfan contaminated water.

In all body tissues studied, the level of damages are more in acute than in sublethal exposure.

Recently shellfish fisheries has developed to such an extent that it plays a significant role in the Indian Economy. There is vast scope in culturing in the back water and estuaries along the Indian coast. Clams are nutritious, delicious and abundant sources of protein. It paly a significant role in culinary of coastal people especially during monsoon when the fishes are not readily available. They can be cultured by simple and inexpensive method. Tellinids are consumed by people and it is a good field to fihses too. The knowledge of biology, breeding habits and nutritive values are essential for shellfish culture. Baseline information collected in the present work on the biology of Tellinids, spawning periods, breeding habits, physico-chemical parameters and nutritive value from Porto Novo waters can be fruitfully utilized in culture practices.

Widespread and indiscriminate use of the pesticides in agriculture has been causing a great concern as the environment is being dangeroulsy contaminated. The physiological activities like feeding, respiratory mechanism, excretion and various life activities such as ciliary activity, burrowing, byssal activity and also growth have been seriously impaired on exposure to pollutants even at low concentration. In the present investigation it has been observed that the organs like siphon, ctenidium, foot, digestive diverticula and gonad (oocyte) are seriously affected in both acute and sublethal concentrations. The organs that are affected functionaly performed to hold vital functions such as respiration, feeding, locomotion and reproduction. Since the sensitive and storage organ digestive diverticula is also affected, cautious application of pesticides are unavoidable. Further work regarding replaceable pesticides and determination of MAC (Maximum allowable concentration) of pesticides will throw more light on successful aquaculture.

APPENDICES

PROCEEDINGS OF THE FIFTH INDIAN SYMPOSIUM OF INVERTEBRATE REPRODUCTION 1987

Changes in the bio-chemical composition of *Conus amadis* (Gmelin) (Mollusca : Gastrpoda) during development

T. SARADHA, J. TAGORE and R. KASINATHAN

Centre of Advanced Study in Marine Biology, Annamali University, Parangipettai 608 502 Tamil Nadu, India.

ABSTRACT

Biochemical composition (protein, fat and carbohydrate) during the development of **Conus amadis** has been studied in freshly spawned, one celled, two celled, four celled, multicellular stage and first to fourth stage of veliger. All stages contained low carbohydrate. Protein content was higher than that of lipid. Lipid and carbohydrate were ultized during early development In adult protein was the major biochemical constituent Biochemical composition of the female was higher than that of males.

1. Introduction

The genus *Conus* is confined to tropics associated with coral reefs. The animals are typically nocturnal, burrowing in sand under stones (Kohn 1958; Halasted 1965). Studies on Conidae are few from Indian waters, but all studies focussed to taxonomy and toxicology. Reports are available on the toxins of *Conus* and their effects on animals from Portonovo waters (Kasinathan *et al* 1987). The literature on reproduction and larval development of tropical marine prosobranchs has been greatly accumulated (tucker Abbott 1954; Natarajan 1957). Yolk utilization during embryogenesis has been studied in a variety of vertebrates and invertebrate groups (Needham, 1966). Since there is paucity of information during the development of *Conus amadis* from portono waters, the present study has been undertaken.

2. Materials and Methods

The egg capsules of *C.amadis* were identified and transferred from the shore to the laboratory. They were maintained in finger bowls containing natural medium. The medium was frequently changed and the development was followed. The biochemical composition of developing embryos has been examined in the following stages: Freshly spawned, one celled, four celled, multi cellular and first, second, third and fourth stages of viliger. On reaching a particular stage, embryos were taken for biochemical analysis. Different stages were then dried for 24 hr at 105°C. It was not possible to weigh individual embryo and hence groups of at least 50 in number in each stage were counted, weighed and dried. The dried embryos were finely powdered for estimating biochemical contents. Protein was estimated by Biuret method (Raymont *et al* 1964); carbohydrate by methanol method (Folch *et al* 1956) and lipid by chloroform-methanol method (Folch *et al* 1956) and then analysed in the Hitachi double beam spectrophotometer (Model 12208), Adult *C. amadis* in different size groups of both the saxes (ranged between 25-30, 30-35, 35-40, 40-45 and 45-50mm) were also taken and the same procedure was followed for biochemical analysis and the results were expressed in percentage.

3. Results

Values for protein, lipid and carbohydrate contents of developing stages are given in Figs 1, 2 and 3. In general, all stages of *C. amadis* contained low carbohydrate. Protein content was higher than that of lipid. Embryo utilized significant amount of lipid for development through each stage. The lipid ranged from 26.6 to 28% during development. Protein content increased during later stages of development and reached 43.4%. The carbohydrate ranged from 1 to 2.4%. In contrast to the utilizatim of protein, lipid and carbohydrate were spent during early developmental stages. Fig. 4 shows that the percentage of biochemical constituents in the adult was higher than the embryo. In adult *C. amadis* the protein was the main constituent as found in earlier stages. Protein ranged from 50.4 to 54.2%; lipid ranged from 24.4 to 26.6% and the carbohydrate from 1.7 to 4.3%

4. Discussion

In term of biochemical composition the gastropod *C. amadis* embryo adheres to the typical molluscan pattern of development. The protein is the main constituent followed by lipid and carbohydrate. In all these stages protein ranges from 41.8% to 43.4%. These quantities of protein are comparable to the values during the development of the gastropod *Crepidula fornicata* (Pandian 1969) where the protein content relatively unchanged. Protein metabolism is restricted mostly to planktonic marine embryos and demersal eggs. The fat content of *C. fornicata* shows a remarkable

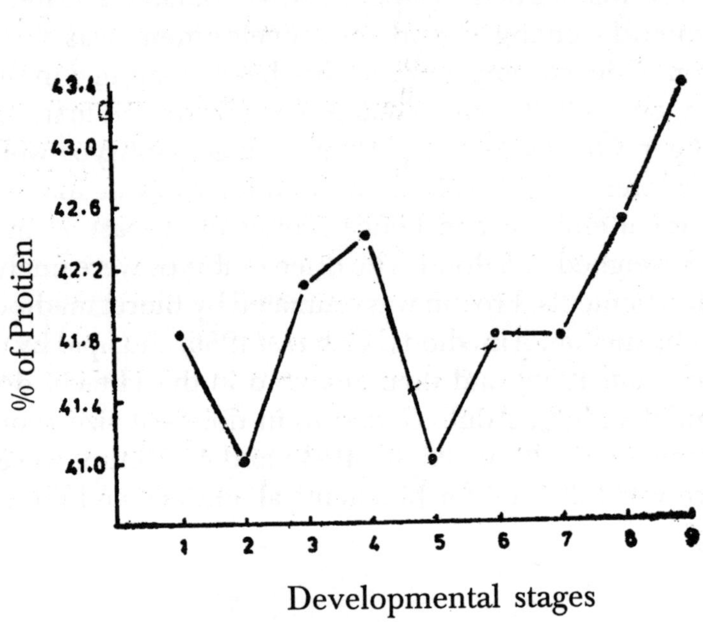

Fig. 26 : Changes of protein during the embryonic stages of *C. amadis.*

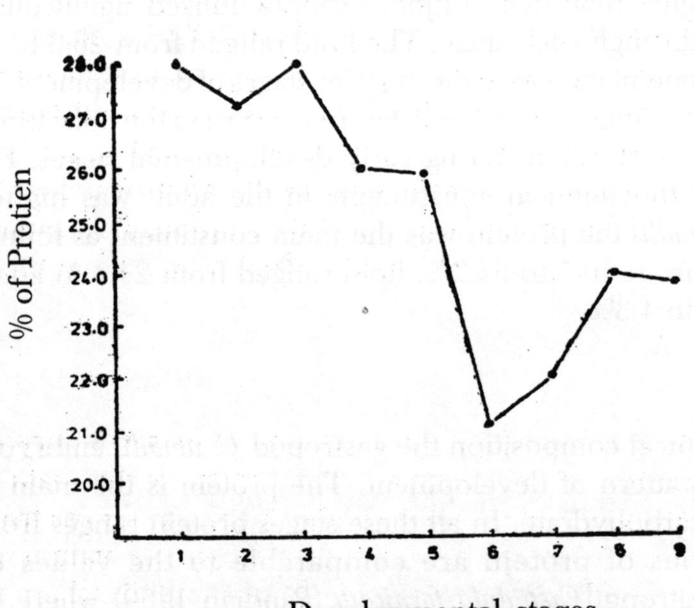

Fig. 27 : Changes of lipid during the embryonic stages of *C. amadis.*

decrease from 33.7% in fresh egg to 20.3% in the veliger. The present result also shows a variation of lipid which ranged from 23.6 to 28%

Needham (1966) identified two patterns of yolk utiliztaion in oviparious organisms. In terrestrial eggs energy for development is obtained primarily by lipid metabolism, which gives much metabolic water and non-toxic waste products. On the other hand, protein provides the most of the energy for the development in aquatic egg, which are not confronted with problems of dehydration and waste disposal of magnitudes experienced by terrestrial egg. Accordingly *C. amadis* follows typical aquatic development. Changes in carbohydrate are small during embryonic development carbohydrate content is decreased from 2.4 to 0.93% in the selected stages. Perhaps a portion of this carbohydrate may be contributed to structural formation. In early embryonic development of *Euphasia superba* (Antartic krill) carbohydrates are utilized at a significant rate (Margaret *et al.* 1984). Yet carbohyrates are minor energy source in embryos and are usually present in low amounts (Holand 1978).

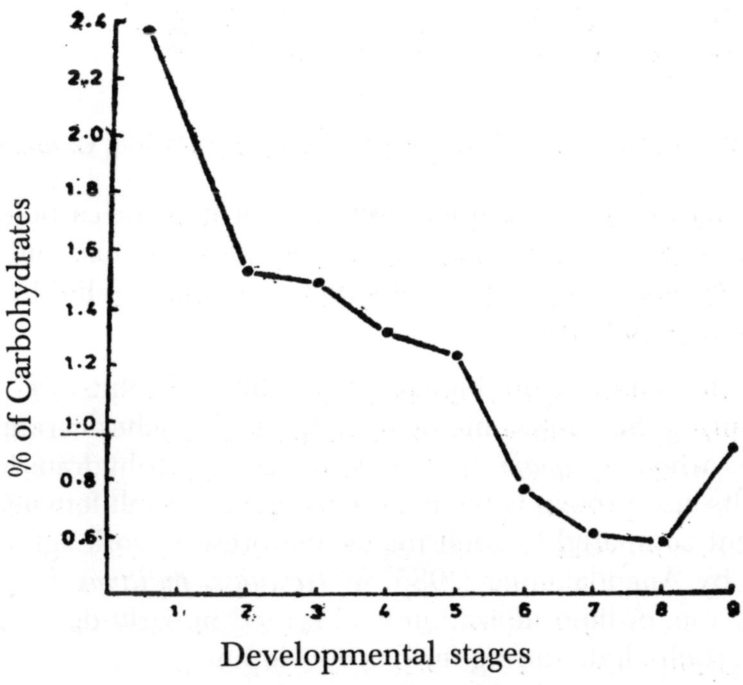

Fig. 28 : Changes of carbohydrate during the embryonic stages of *C. amadis*. 1. One celled stage 2. Two celled stage 3. Four celled stage 4. Eight celled stage 5. Multi – cellular stage 6. I Veliger stage 7. II Veliger stage 8. III Veliger stage 9. IV Veliger stage.

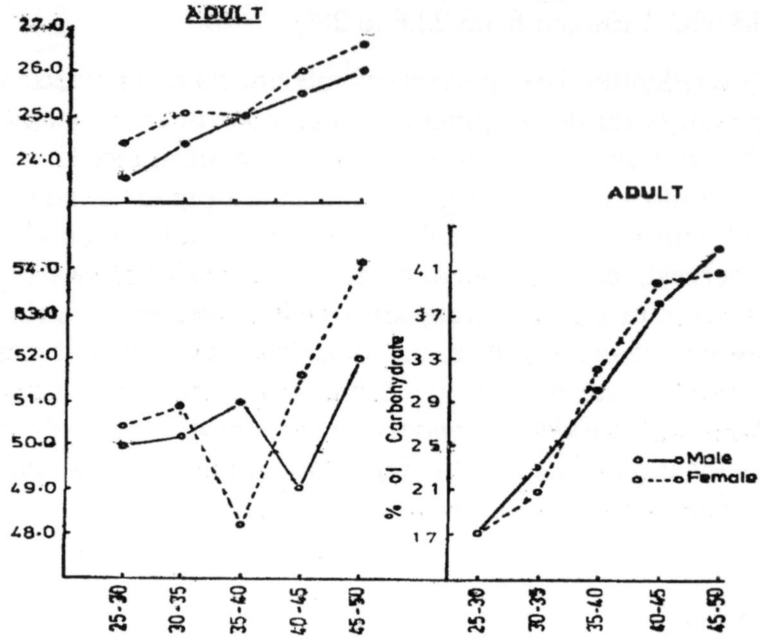

Fig. 29 : Effects of size on protein carbohydrate and lipid of adult *C. amadis*

There are meagre changes in the proportional utilization of lipid and carbohydrate components. Between the freshly spawned one cell to veliger, 5.4% of lipid and 1.54% of carbohydrate are lost. During embryonic development one-third of lipid content was lost when compared to carbohydrate.

When compared to the embryos, the biochemical values of adults are always higher. Lipid decreases during the embryonic development. It reaches certain limit and thereafter it increases when *C. amadis* increases in size. Carbohydrate level is slightly increased. In adults also protein is the main constituent. Adult females have higher biochemical content compared to adult males. the present work agrees well with the values obtained by Anandakumar (1985) in *Hemifuses pujilinus*. It may be concluded that the steady rate of lipid utilization is observed in early development besides the protein and carbohydrate during further development.

Acknowledgements

The authors are grateful to prof. V.K. Venugopalan for his keen interest and encouragement. Grateful thanks are due to the ICAR and UGC New Delhi for the financial support.

REFERENCES

Anandakumar, S., 1985, studies on *Hemifusus pugilinus* (Born) (Mollusca: Gastropoda: Volemidae) from Portonovo water; M. Phil Thesis Annamalai University; g

Dubois, M. Giller, K. A. Hamilton, J. K., Robors, R. A. and Smith, F., 1956, Colorimetric method for determination of sugars related substances; *Analyt. chem* 28: 350-56.

Folch, J. Lees M. & Sloane stanley, G. H. ,1956, A simple method for the isolation and purification of total lipids from animal tissues: *J Biol Chem 226* 497-509.

Halstead, B. W., 1965, Poisonous and venomous marine animals of the world. United States Government Printing Office; 1.

Holand, D. L., 1978, Lipid reserves and energy metabolism in larvae of benthic invertebrates In: Mallins D. C, Sargent J. R. (eds) Biochemical and bio-physical perspectives in marine biology, Academic Press London ; 4 : 85-123.

Kasinathan, R., Shanmugam, A. and Tagore, J., 1987, A preliminary report on venom apparatus in *conus amadis* (Gmelin) from Protonovo coast; (*Curr. Sci.*) 20-190-91.

Kohn, A. J., 1958, Recent cases of human injury due to venomous marine snails of the genus *Conus. Hawaii Msd Jour.* 7 : 1528-32.

Margaret, O. Leary Amsler and George, R. Y., 1984, Seasonal variation in the Biochemical composition of the embryos of *Callinectes sapidus* (Rathbun); *Jour. Crust. Biol.* 4 546-63.

Natarajan, A. V., 1957, Studies on the egg masses and larval development of some prosobranchs from the gulf of Mannar and Palk Bay : *Proc. Indian Acad. Sci.* 46 170-228.

Needham, J., 1966, Biochemistry and morphogenesis : Cambridge University Press London ; p. 787.

Pandian, T. J., 1969, Yolk utilization in the gastropod *Crepidula fornicata : Mar Biol.* : 117-21.

Raymont, J. E., Gaustin, A. and Lingform, E., 1964, Biochemical composition of *Neomysis integer; J cons perm Explor Mier.* 28 : 354-63.

Tucker Abbott, 1954, American sea shells. Van Nostrand company. INC Princeton New Jersey : 1 : 24-29.

PROCEEDINGS OF THE FIFTH INDIAN SYMPOSIUM OF INVERTEBRATE REPORUDCTION 1987

Effect of dimethoate on bio-chemical constituents of reproductive tissues of *Odontopus varicornis* (Dist. (Hemiptera) : Pyrrhocoridae)

M. PREMANATHAN, J. TAGORE, T. SARADHA and R. KASINATHAN

Centre of Advanced Study in Marine Biology Annamalai University
Parangipettai 608 502 Tamil Nadu, India.

ABSTRACT

Pathological effects of the organophosphorus insecticide dimethoate (Rogor) on the reproductive tissues of the adult *Odontopus varicornis* have been examined using biochemical techniques. The insect when injected with 25 μl of the LD50 dose of dimethoate showed reduction in biochemical composition after 72 hours. *O. varicornis* were subjected to different sub-lethal doses. Reduction in the biochemical constituents was observed in L. D. treated and sub-lethal levels.

1. Introduction

The recent advances on the use of insecticides for pest management have raised numerous questions which are yet to be solved. Observations on the injurious effects of pesticides on non-target organisms have been made extensively on different s, stems (Moriarity 1969). A number of authors have also investigated on the histopathological effects of insecticides on reproductive tissues of insects (Atallah and Newson 1966; Rajinder *et al* 1965; Kennedy 1947). The application of certain other chemicals such as chemosterilant, plant extract, juvenile harmonic analogs, fatty acids, the orders such as Coleoptera (Saxena and Vinod Kumar 1982), Hemiptera (Prabhu and Jalaja 1975) and Diptera (Morgan 1967). Recently Saradha (1985) has described the histopathological effects of dimethoate on ovarian tissues of *O. varicornis.* In view of paucity of information on the effects of dimethoate on biochemical contents of reproductive tissues of *O. varicornis,* present study is aimed to describe the effects in the reproductive tissue of *O. varicornis.*

2. Materials and Methods.

Adult *O. varicornis* were collected from the fields in the vicinity of Annamalainagar, Tamil Nadu. They were reared in wooden cages in the laboratory at 29 ± 1°C. It is known that the application of high concentration (lethal concentration) of pesticide resulted in 100% mortality at 24 hrs. Similarly treatment with a low concentration (sub-lethal concentration) of the pesticide produced some mortality at 72 hrs. The median lethal dosage was derived from the nomograms (Litchfield and Wilcoxon 1949). Adult male and female insects were injected with 0.025ml (25 μl) of the lethal and sublethal dose of dimethoate using a hypodermic syringe mounted with 26 gauged sterilized insulin needle through intersegmental region. The control individuals were injected with 0.025 ml of distilled water. The injection site was sealed with a low melting wax following the application of the antibiotic to prevent the loss of haemolymph.

After 3 days of treatment of reproductive tissues of treated and control insects were dissected out for biochemical analysis. Tissues were ovendried for 24 hrs at 105°C. It was not possible to weigh the tissues of single insect and hence a group of at least 10 gonads in each experiment was counted, weighed and dried. The dried tissues were finely powdered for biochemical analysis. Protein was estimated by Biuret method (Raymont *et al* 1954) using bovine serum albumin as standard. Carbohydrate was estimated by sulphuric acid-phenol method. (Dubois (*et al* 1956) using glucose as standard and lipid by chloroform ethanol method (Folch *et al* 1956) and analysed in the Hitachi double beam spectrophotometer Model (12208).

3. Results

Values for the biochemical contents in the reproductive tissues of control and treated (lethal and sublethal) *O. varicornis* were given in Tables 1 and 2. A significant reduction in the percentage of chemical composition was observed in treated insects compared to control. Protein content was higher than that of lipids and carbohydrates. Protein content in male *O. varicornis* which was 19.36% decreased to 10.6% and 15.44% in lethal and sublethal treated *O. Varicornis*. The percentage of lipid in control male and female insects were 4.2% and 5%. It was decreased to 3.9% in lethal dose and 4.15% in sublethal dose in male individual and in ovary, it was reduced as 4.29% and 4.9% respectively. The level of carbohydrate was lower than protein and lipid, both control and insecticide treated insects.

Table 1. Effect of dimethoate (Lethal dose 0.01) on the biochemical composition in the reproductive tissues of the adult *O. varicornis*

Dose	Experiments	Protein (%)	Lipid (%)	Carbohydrate (%)
Lethal dose (0.01)	Control male	19.36	4.23	0.6
"	Treated male	10.6	3.9	0.4
"	Control female	21.66	5.0	0.7
"	Treated female	10.6	4.29	0.45

Table 2. Effect of dimethoate (sublethal dose 0.001) on the biochemical composition in the gonads of the adult *O. varicornis*

Dose	Experiments	Protein (%)	Lipid (%)	Carbohydrate (%)
Sublethal dose (0.001)	Control male	19.36	4.23	0.6
"	Treated male	14.52	4.15	0.45
	Control female	21.66	5.0	0.7
	Treated female	15.44	4.9	0.5

4. Discussion

The insecticide dimethoate affected the biochemical constituents of both tests and ovaries of *O. varicornis.* Changes in biochemical constituents have some correlation with histopathological effects. (Saradha 1985). Dimethoate treated *O. varicornis* exhibited reduction in oviposition, hatching of the eggs, inhibition of ovarian development and destruction of reproductive tissues (Ramalingam 1985). After exposure to various insecticides many insects exhibit decreased egg laying, sterile egg, premature or defective births or hatching difficulty in metamorphosis and reduction in biochemical constituents (Steinhaus 1949). Sometimes the ill effects appear to be carried over to the second generation.

Observations on the effect of pesticide have been made extensively on different system such as digestive system, nervous system, excretery, neuroendocrine system and reproductive system. It is evident from a review of literature that some works have been done on the histopathological effects of insecticides on overian tissues in insects. Many investigations have established that chemosterilants and insecticides produce cytological, physiologicla and genetic effects. The investigations of Morgan and LaBreoque (1965), Morgan (1967) and Rezabova (1968) have shown that chemosterilants act through different biochemical pathways resulting in the inhibition of overian development. The present result shows reduction in biochemical composition in gonads compared to the control insects due to insecticide treatment.

Acknowledgement

The authors are greateful to Prof V. K. Venugopalan for his keen interest and encouragement. Thanks are due to the UGC and ICAR, New Delhi, for the financial support.

REFERENCES

Atallah, Y. H. and L. D. Newson, 1966, Ecological and nutritional studies on *Colemigilla maculata* De Geev (Coleoptera) III. The effect of DDT, toxaphene and Endrin on the reproductive and survival potentials. *J. Econ. Ent.* 59 : 1181 –1187.

Dubois, M. D. A. Giller, J. K. Hamilton, R. A Rebors and F. Smith, 1956, Calorimetric method for determination of sugars and related substances. *Analyt. Chem.* 28 : 350-56.

Folch, J. M. Lees and G. H. Sloane Stanley, 1956, A simple method for the isolation and purification of total lipids from animal tissues. *J. Biol. Chem.* 226 : 497 – 509.

Kennedy, T. S., 1947, The excitant and repellent effects on mosquitoes of sublethal contacts with DDT. *Bull. Ent. Res.* 37 : 593-607.

Litchfield, J. T. and F. Wilcoxan, 1949, A simplified method of evaluating dose-effect experiments. *J. Pharmacol. Exper. Ther.* 96 : 99-113.

Morgan, P. B., 1967, Effects of hempa on the ovarian development of hoseflies, *Musca domestica* L (Diptera : Muscidae). *Ann. Ent. Soc. Amer.* 60 : 812-828.

Moriarity, F., 1969, The sublethal effects of synthetic insecticides on insects. *Biol. Rev.* 44 : 321-351.

Prabhu, V. K. K. and M. Jalaja, 1976, Effect of chemosterilants apholate and metepa on the ovaries of the red cotton bug. D. *cingulatus* (Heteroptera : Pyrrhocoridae). *Entomon* 1 : 45-53.

Rajinder, S. Saini and L. K. Cut-Komp, 1965, The effects of DDT and sublethal doses of Dicofol on reproduction of the two spotted spiders, *J. Econ. Ent.* 49 : 249-254.

Raymont, J. E. G., A. Austin and E. Lingford, 1964, Biochemical studies on marine zooplankton. 1. Biochemical composition of *Neomysis integer*. *J. Cons. Perm. Explor. Mer.* 28 : 354-63.

Rezabova, B., 1968, Changes in the metabolism of nucleic acids in the ovaries of the housefly, *Musca domestica* after application of chemosterilants. *Acta Entomol. Bohemoslov.* 65 : 337 – 340.

Saradha, T., 1985, Studies on the effects of the insecticide Dimethoate on ovarian tissues in *Odontopus varicornis* (Dist) Hemiptera : Pyrrhocoridae). M. Phil. dissertation. Annamalai University.

Steinhaus, A., 1949, *Principles of insect pathology;* McGraw Hill Publication. 36-38.

Saxena, S. C. and Vinod Kumar, 1982, Effect of two chitin inhibitors on reproduction of *Trogoderma granarium*. *Entomon* 7 (2) : 141-144.

PROCEEDINGS OF THE FIFTH INDIAN SYMPOSIUM OF INVERTEBRATE REPORUDCTION, 1987

Studies on metallic pollution in the gonads of bivalves from Porto Novo Waters

J. TAGORE, A. SHANMUGAM, T. SARADHA and R. KASINATHAN

Centre of Advanced Study in Marine Biology,
Porto Novo 608 502 Tamil Nadu, India

ABSTRACT

Gonads of both the sexes of *Sunetta scripta* from Porto Novo coast, *Tellina nobilis* from Vellar estuary and *Crossostrea madrasensis* from Killai backwaters were collected for one year for trace metal analysis. Analysis of the gonads of these animals showed seasonal variations in the concentrations of copper and zinc. The results indicate the impact of metalic concentration on the gonads of these bivalves.

1. Introduction

Increasing awareness of the potential hazards of contamination of the ecosystem with toxic elements from urban sources has high lighted the need for continous monitoring of the levels of these substances in the environment. Among Pollutants trace metals are important that affect aquatic fauna. The trace metal content in an intertidal oganism could either increase or decrease with reponse to the trace metals in the overlying water and sediment (Tagore *et al* 1987). Bivalves have received extensive treatement in the literature owing to their reported ability to reflect environmental levels 1980; Phillips 1977; Langston 1982). Accumulation of metals in bivalves has been observed in Porto Novo waters (Kumaraguru 1980; Mahondran 1985 : Senthilnathan *et al* 1986). In all of these works the authors have focussed only on the whole body of the organisms. The gonads which are one of the sites of heavy metal accumulation are important than any other organs. Yet information concerning the concentration and its effects of the concentration of metals in the gonads of selected edible bivalves.

2. Materials and methods

The bivalves were collected at monthly intervals from January to December 1986 from South east coast of India. The venerid clam *Sunetta Scripta* (Linnaeus) from Porto Novo coast, edible oyster *Crassostrea madrasensis* (Preston) from Killai backwaters and the paper shell *Tellina nobilis* (Hanly) from mouth of the Vellar estuary were collected. Samples in sufficient number of relatively standard sized organisms were collected. In average shell length of S. *Scripta C. madrasensis* and *T. nobilis* were 3.5-4.00 mm, 7.5-8.0 mm and 4.5-5.0 mm, respectively. Animals were collected from their natural beds; after washing with distilled water, the gonads were taken in both the sexes. They were labelled and dried in an oven at 100°C for 24 hr. The metals were estimated in the dried in an oven at 100°C for 24 hr. The metals were estimated in the dried material after wet digestion with perchloric acid and nitric acid. The finely powdered meterial was digested and made upto a constant volume with 0.5 N HNO_3. Trace metals (copper and zinc) were estimated by atomic absorption spectrophotometric method using a varian Techtron AAS, Model 1100 in an Air Acetylene flame.

3. Results and Discussion

Among the metals studied, the accumulation of zinc was the highest in the gonads during the periods of investigation (Fig. 1, 2 and 3). Levels of zinc and copper were comparable in magnitude in these species. Zinc was accumulated in higher concentrations in *C. madrasensis* during August (3.30ppm) and lowest concentration in *S. scripta* during June (4.1 ppm). Based on the results copper showed a distinct fluctuation, recording high values in *C. madrasensis* during August (162 ppm) and low level was registered in *T. nobilis* during March (17.5 ppm)

In highly polluted waters it is unlikely that the heavy metal content of bivalves will reflect the dissolved concentration of those heavy metals which constitute only a small fraction on the total dissolved load. From the results the gonadial levels of zinc in selected animals is always greater than the coppoer concentrations. The accumulation capacity of female is always greater than males; similar results have been reported in *Hemifusus pugilinus* (Anandakumar 1985) and *Crassostrea madrasensis* of Cuddalore backwaters (Rajendran *et al* 1987). Trace metal concentrations of copper and zinc are more in the gonads of edible oysters than other benthic bivalves. Rajendra *et al* (1987) indicated that increased iron content on *C. madrasensis* tissues provokes a relative decrease of the zinc and copper. Rajendran and Kurian (1986) used oysters as an indicator of heavy metal polluion in Cochin backwaters.

Accumulation of metals showed a clear seasonal variation. Metals are higher in pre-monsoon and monsoon seasons than in summer season, which might be due to the inflow of freshwater load with copper and zinc. Subramanian (1981) also reported decreased concentration of metals during summer in the oyster from Pitchavaram mangroves. Bryan (1973) suggested seasonal factors are important governing the availability of metals in the environment and in turn to the organisms.

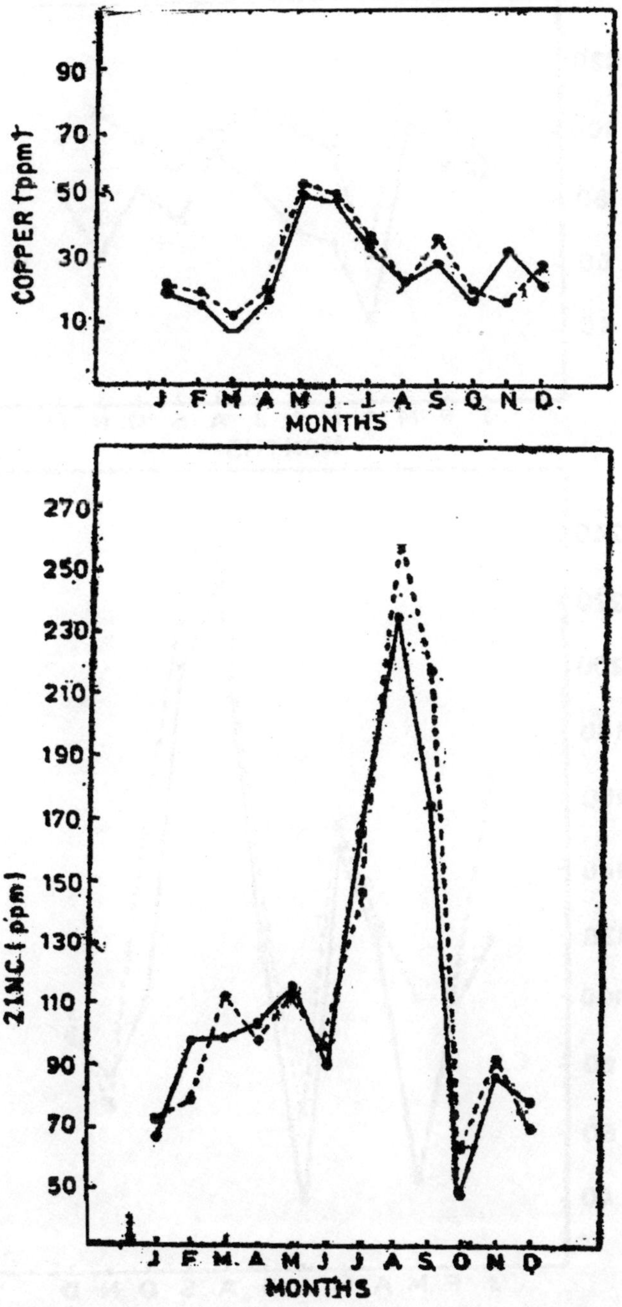

Fig. 30 : Metallic pollution in the gonads of *Tellina nobilis.*

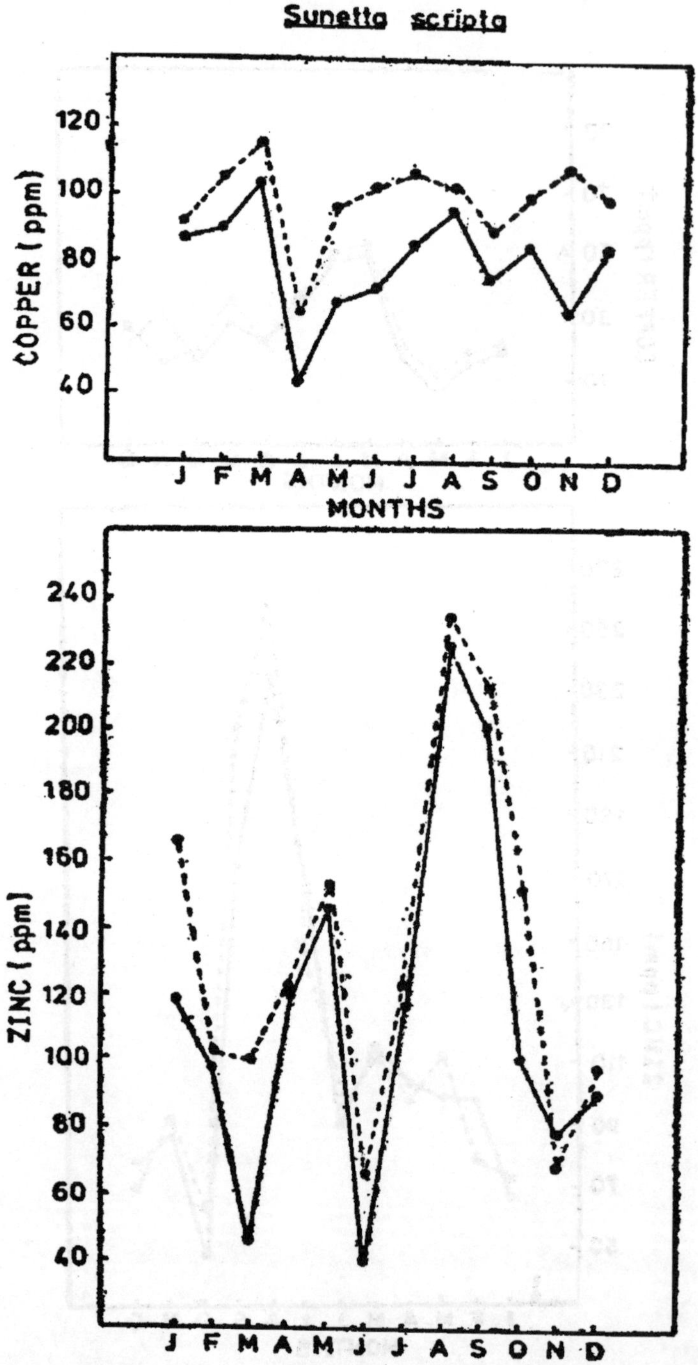

———— Male, ...Female : for figures 30, 31 and 32

Fig. 31 : Metallic pollution in the gonads of *Sunetta scripta*.

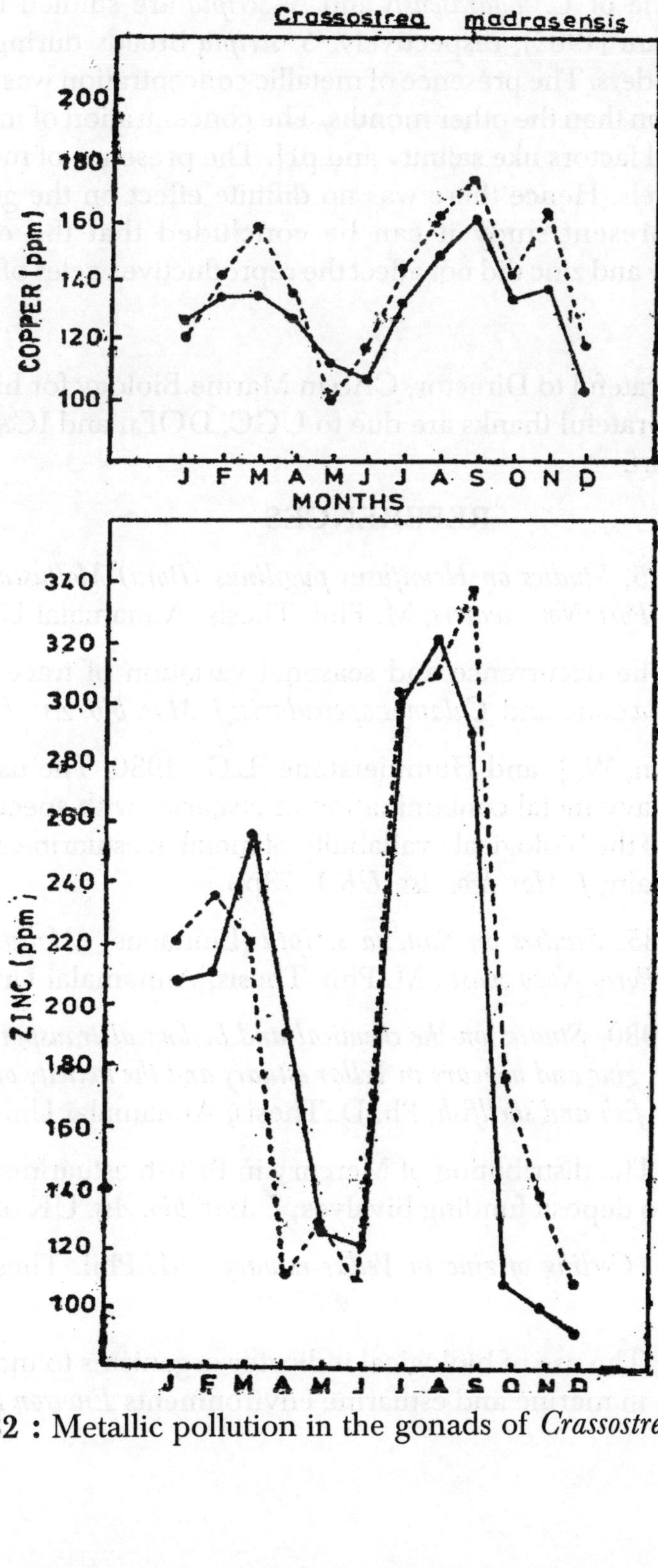

Fig. 32 : Metallic pollution in the gonads of *Crassostrea madrasensis.*

Reproductive cycle of *C. madrasensis* and *S. scripta* are studied by Marimuthu (1985) and Dravidamani (1985), respectively. *S. scripta* breeds during summer, but oysters are annual breeders. The presence of metallic concentration was comparatively lower in breeding season than the other months. The concentration of metals in gonads depends upon seasonal factors like salinity and pH. The presence of metals in gonads was in permissible levels. Hence there was no difinite effect on the gonads of these bivalves. From the present study it can be concluded that the environmental contaminants of copper and zinc did not affect the reproductive cycles of these bivalves.

Acknowledgements

The authors are grateful to Director, CAS in Marine Biology for his keen interest and encouragement. Grateful thanks are due to UGC, DOEn and ICAR, New Delhi for the financial support.

REFERENCES

Anandakumar, S., 1985, *Studies on Hemifusus pugilinus (Born) Mollusca : Gastropoda : Volemidae) from PortoNovo waters;* M. Phil. Thesis, Annamalai University.

Bryan, G. W., 1973, The occurrence and seasonal variation of trace metals on the scallop *Pecton maximus* and *Chlamys opercularis: J. Mar. bio. Ass. UK* 53: 145-66.

Bryan, G. W., Langston, W. J. and Hummerstone. L.G., 1980, The use of biological indicators of heavy metal contamination in etsuaries with special reference to an assessment of the biological availability of metal in estuarine sediments from South West Britain; *J. Mar. bio. Ass. UK* 1: 73pp.

Dravidamani, M., 1985, *Studies on Sunetta scripta* (Linnaeus) (*Mollusca : Bivalvia : Veneridae) from Porto Novo coast :* M. Phil. Thesis, Annamalai University.

Kumaraguru, A. K., 1980, *Studies on the chemical and biological transport of heavy metal pollutants copper, zinc and mercury in Vellar estuary and the toxicity of three pollutants to some estuarine fish and shellfish;* Ph. D. Thesis, Annamalai University.

Langston, W. J., 1982, The distribution of Mercury in British estuarine sediments and its availability to deposit funding bivalves; *J. Mar. bio. Ass.* UK 62: 667-84.

Mahendran, A., 1985, *Cycling of zinc in Vellar estuary :* M. Phil. Thesis, Annamalai University.

Phillips, D. J. H., 1977, The use of biological indicator organisms to monitor the trace metal pollution in marine and estuarine environments *Environ Pollut,* 13 : 281-317

Rajendran, N. and Kurian, C. V., 1986, *Crassostrea madrasensis* (Preston). Indicator of metal pollution in Cochin backwaters. *Proc. National Seminar on Mussel Watch* 1 : 120.32.

Rajendran, N., Tagore, J. and Kasinathan, R., 1987, Heavy metal concentration on oyster *Crassostrea madrasensis* (Preston) of Cuddalore backwaters; *Indiarn J. Mar. Sci.* (In Press).

Shanthi, B., Rajan, A. and Kalyani, M., 1986, Variation on iron concentration in different soft body parts of the claim *Meretrix casta* in relation to season, weight, sex and reproduction *Proc. National Seminar on Mussel watch;* 1: 183-84.

Senthilnathan, S., Mahendran, A. and Balasubramanian, T., 1986, Bioconcentration of iron in *Crassostrea madrasensis* (Preston) from Vellar estuary. *Proc. National Seminar on Mussel Watch* 1 : 160:68.

Subramanian, A. N., 1981, *Some aspects of iron, manganese, copper, zinc and phosphorous in Pitchavaram mangrove;* Ph. D. Thesis, Annamalai University.

Tagore, J., Shanmugam, S. and Kasinathan, R., 1987, Bioaccumulation of iron and manganese in *Sunetta scripta* (Linnaeus) from Porto Novo coast. National Seminar on Ectoxicol, Aurangabad.

PROCEEDINGS OF THE SECOND NATIONAL SYMPOSIUM ON ECOTOXICOLOGY 9-11 JUNE, 1988

EFFECT OF ENDOSULFAN ON THE RATE OF FILTRATION AND EXCRETION OF VENERID CLAIM *Sunetta scripta* (Mollusca : Bivalvia : Vineridae).

T. SARADHA, J. TAGORE, P. MADESWARAN and R. KASINATHAN,

Centre of Advanced Study in Marine Biology,
Portonovo 608 502, Tamil Nadu, India.

ABSTRACT

The venerid claim *Sunetta scripta* from intertidal areas of Portonovo coast was exposed to endosulfan. The present results revealed that filtration mechanism and ammonia excretion in the *S. scripta* impaired due to the effect of endosulfan. In control the animal activity filtered the suspended material and ejected the pseudofaeces. Due to pesticide treatment very low pumping rate and pseudofaeces production were observed in different sublethal levels when compared to control. The level of ammonia decrease in ammonia excretion of the clam under stress.

INTRODUCTION

Much investigations have been carried out in the venerid claim *Sunetta scripta* with respect to biology bioaccumulationof heavymetals[13] and its biochemistry[11]. There is very little information available pertaining to the action of pesticides[6]. Hence efforts have been made to study the effect of pesticide endosulfan on the physiological activities of the clam. S. *scripta.*

MATERIAL AND METHODS

The bivalves *S. scripta* were collected from Portonovo coast and were brought to the laboratory, cleaned and acclimated (7, 5) and salinity (20 %) was maintained constant throughout the experiment. The clams with an extended siphon and valves opened sufficiently to display the edge of the mantle were selected. The toxicity tests were conducted in accordance with the methods recommended by the committee on methods for toxicity test with aquatic for each set of experiments. The bioassays were terminated at the end of the test period (7 days). The mortality rates and behaviour of the bivalves in each test concentrations were recorded at regular intervals. The experiment was repeated thrice and the values of LC_{50} (60 ppb) and different sublethal levels (30 ppb and 10 ppb) were determined by using probit analysis.

Table 1 : Effect-of endosulfan on rate of filtration of *Senetta scripta*

Sl. No.	*Experiment*	*% of rate of filtration*			
		15 mts.	*30 mts.*	*45mts.*	*60mts*
1.	Control	39.0	40.9	41.5	42.8
2.	LC_{50} 60 ppb	5.5	4.5	4.0	4.0
3.	30 ppb (sublethals)	5.6	5.4	5.2	5.0
4.	10 ppb	2.3	20.0	20.0	21.0
5.	2 ppb	40.0	44.2	46.1	47.0

Table 2: Estimation of ammonia (NH_4) excretion (M_m of NH_4/100 ml of H_2O exposed to endosulfan

Sl. No.	*Experimental concentration*	*12 hr interval*	*24 hr interval*
1.	Control	0.4382	0.4420
2.	60 ppb	0.3528	0.2971
3.	10 ppb	0.4120	0.3976
4.	2 ppb	0.4500	0.4261

The shells were placed in 400 ml filtered estuarine water separately and allowed to recover from handling for 10-20 minutes. 4 ml of neutral red were added to each test vessel and gently mixed by stirring. The dye solution was made by dissolving a quantity of solid dye in a little distilled water. Samples of 10 ml were immediately extracted with a pipette from each vessel. At 15 minutes interval 10 ml of solution was removed from each container during the course of 1 hr. Dye concentration in the samples were determined spectrophotometrically at 425 nm in a Hitachi double beam spectrophotometer after the addition of a few drops of hydrochloric acid. The technique followed was the same as was used by Cole and Hepper[2] and Nagabhushanam[8]. On each concentration about 10 clams were used. The readings were converted into per cent concentration by the use of a calibration curve. The rate of ammonia excretion was measured by using Nessler's reagent[10] at 12 hr. intervals.

RESULTS

In control the venerid clams filtered the neutral red, suspended particles and ejected the pseudofaeces actively. But these physiological activities were found to be impared in endosulfant treated clams. When compared to control, in the acute concentration (60 ppb) the animal exhibiting reduced activity and also the pumping rate and the pseudofaeces production were very low. The present result revealed that the efficiency of filtering capacity is reduced from 42% to 4% in the control and experimental animals respectively. At the concentration of 10 ppb the filteration was increased and it reaches half the level of control value (i.e.) 21% (Fig. 1). And also in 10 ppb the animals were active, the pumping rate and the producion of pseudofaeces were higher than the acute and high sublethal levels. At very low concentration of 2 ppb the filtration rate (47 %) was increased than the control animal (Table 1).

Table 2 shows the ammonia content in control as well as endosulfan exposed animals. It has been showed that a considerable drop of ammonia content at all intervals exposed to acute and different sublethal concentrations. The rate of excretion is higher in control (0.4382) than lethal and sublethal levels. Similar to rate of filtration, the excretion rate was found to be increased at lowest sublethal levels (2 ppb).

DISCUSSION

Pesticides cause stress on the organism and disturb its physiological activities. Many authors investigated the toxicity effects on various lamellibranches Davis[3] on *Crassostrea virginica;* Engel *et al.*[5] *on Mercenaria mercenaria;* Abraham *et al.* in *Villorita cyprinoides* and Nagarajan[9] in *Meretrix casta*. Nagarajan reported that pollutant might act as physiological stress for exposed oranisms as do the altered environmental parameters. The gill and labial pulp tisues affected by pesticide endosulfan, may account for decrease in water filtration in increased concentration. In the present investigation it was observed that higher concentration of pesticide seems to arrest physiological activities. At lower concentration the rate of water filtration was slightly increased than the control. The filtration rate once again come down than the control if the experiment was continued.

The results of the present study clearly indicate that the toxicity effects of the pesticide endosulfant have a potent on the physiological activities of the venerid clam

Sunetta scripta. The present results are very similar to the *Meretrix casta* as described by Nagarajan.

ACKNOWLEDGEMENT

The authors are thankful to the Director, C.A.S. in Marine Biology, for providing facilities. Thanks are also due to ICAR and UGC for financial support. And special thanks are also due to Mr. Nagarajan, Assistant Professor for his fullest co-operation in this experiment.

REFERENCES

1. Abraham, T.J., K. Y. M. Salih and J. Chacko, 1986, Effects of heavy metals on the filtration rate of bivalve *Villorita cyprinoids* (Hanley) Indian J. Mar. Sci., 15 : 195-196.

2. Cole, H. A. and Hepper, B. T., 1954, Use of neutral red solution for the comparative study of filtration rates of lamellibranchs. J. Cons. 20: 197-203.

3. Davis, H. C., 1961, Effects of some pesticides on eggs and larvae of oysters (*Crassostrea virginica*) and clams (Venus mercenaria). Commer. Fish. Rev., 23 : 8-23.

4. Dravidamani, M., 1985, Studies on *Sunetta scripta* (Mollusca Bivalvia) from Portonovo coast. M. Phil. Thesis, Annamalai University.

5. Engel, R. H., Neat, M. J. and Hillman, H. J., 1972, Sublethal chronic effects of DDT and Lindane on glycolytic and glycogenic enzymes of the quabog, *Mercenaria mercenaria*, In: Marine pollution and sea life, pp. 257-259, London, Fishing News.

6. Madeswaran, P., J. Tagore, T. Saradha and R. Kasinathan, 1988, Effects of endosulfan on the biochemical constituents of Venerid clam *Sunetta scripta* (Mollusca : Bivalvia) from Portonovo coast. National Symposium on Ecotoxicology, Chidambaram, 9th June, 1988.

7. Mather, K., Fisher, R. A., 1949, Statistical analysis in Biology, Mathran and Co. Ltd., London.

8. Nagabhushanam, R., 1956, The rate of water filtration in the marine wood boring mollusc, *Martesia striata*. Proc. Indian Acad. Sci., 43-223.

9. Nagarajan, 1987, Effect of endosulfan on *Meretrix casta*. M. Phil. Thesis, Annamalai University.

10. Natelson, S., 1971, Techniques in clinical chemistry charles (Springfield, Illinois : Thomson Publishers) 146.

11. Saradha, T., J. Tagore and R. Kasinathan, 1988, Some aspects of biochemistry of the venerid clam *Sunetta scripta* (Mollusca : Bivalvia) paper presented in the National Seminar on Prespectives in Aquatic Biology, 20-22, March, 1988.

12. Sprague, J. B., 1969, Measurement of pollutant toxicity to fish. I. Bioassay methods for acute toxicity. Water Res., 3: 793-821.

13. Tagore, J., A. Shanmugham and R. Kasinathan, 1987, Bio-accumulation of iron and manganese in *Sunetta scripta* (Linnaeus) from Portonovo coast, Proc. Nat. Symp. Ecotoxic. pp. 125-126.

DISTRIBUTION OF HEAVY METALS IN THE EDIBLE MARINE GASTROPODS OF PORTONOVO COAST

J. TAGORE, T. SARADHA, V. S. CHANDRASEKHARAN AND R. KASINATHAN

Centre of Advanced Study in Marine Biology,
Parangipettai 608 502.
Tamil Nadu.

ABSTRACT

Distribution of copper, zinc, iron and manganese were observed in the whole body and soft body tissues such as gill, mantle, digestive diverticula and gonad of the edible marine gastropods *xancus pyrum (*Linnalus) and *Oliva gibbosa* (Born) from the Portonovo coast for a period of six months (June, 1987 to December, 1988). Analysis of tissues demonstrates seasonal variability at all metal levels and the results are discussed accordingly.

INTRODUCTION

Levels of trace metals in large variety of shell fishes would provide baseline data for some of the toxic metals. Most of the current literature on bioaccumulation of heavy metals of molluscs were devoted to the filter feeding bivalve molluscs were studied[2-12]. However, reports in edible gastropods are scant.[1,13] The present paper deals with the distribution and accumulation of trace metals in the marine edible gastropods *Xancus pyrum* and *Oliva gibbosa.*

MATERIAL AND METHODS

The gastropod *Xancuc purum* (Linnalus) from pelagic waters of Portonovo and *Oliva gibbosa* (Born) from intertidal areas of Portonovo coast at montly intervals over a period of six months from June to December, 1987. The animals were washed thoroughly with metal free distilled water. The whole and different soft body tissues such as gill, mantle, digestive diverticula and gonad of both the sexes were pooled, washed and were ovendried at 100°C for 24 hours. The dried tissues were powdered and 0.5 g of each body tissue was digested with perchloric acid and nitric acid[14]. After dry digestion, samples were analysed in AAS (Model, 2380).

RESULTS

The samples were analysed and the data were given in Fig. 1 and 2. The results indicate that in both the species of gastorpods bioaccumulation of iron content was

Table 1 : Bioaccumulation of trace metal contents in *Olive gibbos* and Xancus pyrum (ppm)

Months	*Animal*	*Fe*	*Zn*	*Cu*	*Mn*
Summer					
June	*Oliva gibbosa*	26.5	20.5	18.0	14.5
	Xancus pyrum	22.2	17.5	14.1	13.0
Pre monsoon					
July	*Oliva gibbosa*	12.8	52.5	19.5	19.0
	Xancus pyrum	126.5	51.7	15.8	13.8
August	*Oliva gibbosa*	136.0	70.0	20.0	14.9
	Xancus pyrum	127.0	60.0	17.5	12.1
September	*Oliva gibbosa*	175.0	61.9	28.0	13.8
	Xancus pyrum	170.0	56.0	22.0	11.9
Post monsoon					
October	*Oliva gibbosa*	225.0	107.0	42.0	31.0
	Xancus pyrum	211.0	90.5	39.5	29.1
November	*Oliva gibbosa*	220.0	111.1	44.5	39.4
	Xancus pyrum	217.5	100.0	41.9	37.2
December	*Oliva gibbosa*	201.5	120.0	48.0	55.9
	Xancus pyrum	199.0	110.0	42.5	49.5

Table 2: Bio-accumulation of trace metals in different soft tissues of *Olive gibbosa* (ppm)

Month	*Animal*	*Fe*	*Zn*	*Cu*	*Mn*
June	Gill	147.5	80.0	62.5	41.5
	Mantle	131.0	79.7	59.5	38.9
	Dig. Div.	120.0	52.5	48.5	34.5
	Gonad	121.5	67.5	56.3	35.1
July	Gill	156.5	93.5	87.5	53.0
	Mantle	149.5	92.1	85.9	49.8
	Gonad	132.8	82.5	73.5	33.3
	Dig. Div.	145.8	89.9	79.5	45.2
August	Gill	212.5	87.0	80.5	63.0
	Mantle	211.9	79.8	79.1	61.9
	Gonad	199.5	72.5	69.5	59.8
	Dig. Div.	191.0	69.1	58.9	57.0
September	Gill	290.5	71.5	67.0	48.0
	Mantle	262.7	67.7	58.5	39.9
	Gonad	400.0	52.0	45.5	26.5
	Dig. Div.	409.0	59.8	49.2	27.8
October	Gill	417.5	117.5	117.0	96.0
	Mantle	409.2	110.1	109.1	92.5
	Gonad	400.0	100.8	99.2	90.0
	Dig. Div.	409.0	109.5	100.5	91.5
November	Gill	318.0	120.9	111.9	101.0
	Mantle	311.5	115.0	99.5	99.1
	Gonad	300.0	120.0	100.8	100.0
	Dig. Div.	300.9	137.0	132.8	100.0
December	Gill	412.0	137.0	132.0	67.5
	Mantle	400.1	131.5	127.0	50.9
	Gonad	395.9	129.0	119.5	50.5
	Dig. Div.	399.5	130.0	121.5	58.5

Table 3 : Bio-accumulation of trace metals in different soft tissues of *Xanous pyrum* (ppm)

Month	*Animal*	*Fe*	*Zn*	*Cu*	*Mn*
June	Gill	135.9	75.4	54.5	39.9
	Mantle	132.1	74.2	44.3	34.5
	Gonad	125.0	69.1	39.1	25.7
	Dig. Div.	110.0	51.9	27.5	21.2
July	Gill	225.5	162.7	144.4	131.2
	Mantle	211.0	125.5	124.4	–
	Gonad	119.5	111.0	109.9	107.9
	Dig. Div.	105.1	101.9	100.1	100.0
August	Gill	275.0	179.0	142.0	130.0
	Mantle	225.0	145.4	139.0	125.0
	Gonad	211.2	132.5	121.1	115.0
	Dig. Div.	209.0	111.0	111.5	105.0
September	Gill	299.9	192.5	153.0	145.5
	Mantle	292.5	185.9	152.5	137.9
	Gonad	271.1	171.9	129.0	129.6
	Dig. Div.	260.0	160.0	119.0	120.1
October	Gill	355.0	295.0	189.9	150.5
	Mantle	327.5	294.0	171.0	145.1
	Gonad	317.5	290.0	162.5	130.5
	Dig. Div.	300.0	275.0	155.0	125.0
November	Gill	387.9	298.0	190.0	151.5
	Mantle	317.1	117.0	121.5	101.2
	Gonad	301.8	107.5	111.5	100.9
	Dig. Div.	300.7	100.1	105.5	99.1
December	Gill	390.0	299.0	191.0	152.0
	Mantle	288.5	271.5	190.0	147.9
	Gonad	271.9	260.9	189.5	136.5
	Dig. Div.	216.0	259.0	189.0	132.1

more than other metals. The accumulation was in the following decreasing order Fe> Zn > Cu > Mn. *Oliva gibbosa* acculmulates more content of trace metals than *Xancus pyrum*. It was also observed that high metal loads were found during the monsoon period (October to December) while low metal content was recorded in the summer month (June). Regarding the trace metal accumulation in different soft body tissues, in both the gastropods showed that gill and mantle accumulated more than the other tissues.

DISCUSSION

In the present observation, the bio-concentration of heavy metals in the whole and different soft tissues in the edible gastropods were summarised as follows:

Oliva gibbosa > *Xancus pyrum*

Summer > Premonsoon > Monsoon

Gill > Mantle > Gonad > Digestive diverticula

Iron > Zinc > Copper > Manganese.

From this study it was noticed that accumulation factor concerned with gastropod *Oliva gibbosa* showed greater affinity towards the trace metals than *Xancus pyrum*. It reveals that different marine species repond to different metal load, seems to be not matching with each other in this respect from the same locality. An increase in the trace metal content in the tissues of a benthic mollusc is usually controlled by the increased availability of the trace metal in the bottom sediment[10]. In the present study *O. gibbosa* acculumates more metals than *X. pyrum.* the intimacy of animals with bottom sediment is more in previous than later.

The results showed that maximum concentration of trace metals were recorded during the monsoon season in both the gastrometals were recorded during the monsoon season in both the gastropods and the minimum accumulation was recorded during summer. It was earlier reported that maximum concentration of metals recorded during the end of premonsoon season in oysters and minimum during the post monsoon season.[4] The highest concentration of trace metal encountered during monsoon period, which may be due to heavy rainflow of fresh water rish in metals.

The accumulation of trace metals in different soft tissues of the present edible gastropods showed that gill accumulates more concentration than other tissues. It was reported that gills and mantle are the sites of higher heavy metal concentration due to the presence of mucous which is covering the gills and may be the reason for the high metal content.[2.]

ACKNOWLEDGEMENTS

The authors are thankful to the Director, C.A.S. in Marine Biology, Portonovo for facilities provided and the authors thank to the UGC and ICAR, New Delhi for financial assistance.

REFERENCES

1. Ananda Kumar, 1985, Studies on *Hemifusus pugilinus* (Mollusca: Gastropoda : Volumidae) from Portonovo water, M. Phil, Dissertation, Annamalai University.
2. Kumaraguru, A. K., 1980, Studies on th^e chemical and biological transport of heavy metal pollutants copper, zinc, mercury in Vellar estuary and the toxicity of these pollutants to some estuarine fish and shell fish. Ph. D. Thesis, Annamalai University.
3. Lakshamanan, P. T. and P.N.K. Nambisan, 1983, Seasonal variation in trace metal content in bivalve molluscs, *Villorita cyprinoides* Var. Cochinensis (Hanley), *Meretrix casta* (Chemnitry) and *Perna viridis* (Linnaeus) Ind. Jour. of Mar. Sci. Vol. 12 : 100-103.
4. Magendran, A., 1985, Cycling of zinc in the Vellar Estuary. M. Phil. Dissertation, Annamalai University.
5. Premanathan, M, J., Tagore, R., Kasinathan and K. Kathiresan, 1987, Variation on heavy metal concentration in bivalves collected from Protonovo waters. Proc. Nat. Symp. Ecotoxic, pp. 129-33.
6. Rajendran, N. J., Tagore and R. Kasinathan, 1988, Heavy metal concentration in oyster *Crassostrea madrasensis* (Preston) of Cuddalore back water. Indian J. Mar. Sci. (In Press).
7. Sankara Narayanan, V. N. and K.S. Purushan and T.S.S. Rao, 1978, Concentration of some of the heavy metals in the oyster *Crassostrea madrasensis* (Preston) from the Cochin region. Indian J. Mar. Sci., 7(2): 130-131.
8. Segar, D. A., J. D. Collins and J. P. Riley, 1971, The distribution of the major and minor elements in marine animals Part II. Molluscs. J. Mar. Biol. Ass. U.K., 51: 131-136.
9. Senthilnathan, 1985, Cycling of iron in the Vellar estuary, M. Phil., Dissertation, Annamalai University.
10. Tagore, J. A. Shanmugam and R. Kasinathan, 1987a, Bio accumulation of iron and manganese in *Sunetta scripta* (Linnaeus) from Portonovo coast. Proc. Nat. Symp. Ecotoxic, pp. 125-128.

11. Tagore, J. A., Shanmugam, T. Saradha and R. Kasinathan, 1987b. Metallic concentration in the gonads of bivalves from Portonovo waters. Paper presented in V, All India Symposium on Invertebrate Reproduction. Palani, 24-27th September 1987.

12. Tagore, J. R., Kasinathan, V. S. Chandrasekaran and A. Shanmugam, 1988, Seasonal variations of the heavy metals Cu and Zn in different body parts of a venerid clam *Sunetta scripta* Indian J. Mar. Sci., (In Press.)

13. Thirumavalavane, R., 1987, Studies on *Bobylonia spirata* (Mollusca : Gastropoda : Biccirnodae) from Portonovo coast. M. Phil., Dissertation, Annamalai University.

14. Topping, G., 1973, Heavy metals in shell fish from Scottish waters, Aquaculture vol. 1(4): 379-84.

EFFECT OF ENDOSULFAN ON THE BIOCHEMICAL CONSTITUENTS OF VENERID CLAM *SUNETTA SCRIPTA* (MOLLUSCA : BIVALVIA : VENERIDAE) FROM PORTONOVO COAST.

P. MADESWARAN, J. TAGORE, T. SARADHA and R. KASINATHAN,
CAS in Marine Biology,
Annamalai University,
Portonovo 608 502.

ABSTRACT

The intertidal venerid clam *Sunetta scripta* were exposed to different concentrations of endosulfan. The effects on biochemical constituents like protein, carbohydrate and lipid were estimated after exposing at different sublethal levels. Tissues from whole animal as well as different organs Viz., mantle, gill, digestive diverticula, adductor muscle and gonad were analysed. The present results showed that the pesticide treated clams registered reduced biochemical constituents.

INTRODUCTION

Many synthetic organic pesticides were introduced over 40 years ago, but their effects on aquatic environments are still poorly understood. Frequently, numerous studies evaluating only acute toxicity led to rather broad conclusions concerning pesticides in water.[4] Factors such as temperature, salinity, pH, turbidity and organic content of water may greatly alter the toxcity of a pesticide.[7]

Aquatic animals can accumulate concentrates of certain pesticides many times greater than in water. Organochlorines do not degrade readily, thereby accumulating in tissues in one or more forms. Hence, the pesticide may alter the biochemical constituents of that particular organism.

The aim of the present study was to evaluate the biochemical changes in the edible[1] venerid clam *Sunetta scripta* treated with the technical grade endosulfan, an organochlorine pesticide. Numerous works have been done concerning pesticide effects on biochemical composition of fishes.[9, 10, 14] and crabs[13] as the test organisms but scanty about molluscs[8]. In the present study, an effort has been initiated to determine any possoble changes in the biochemical composition of venerid clam.

MATERIAL AND METHODS

Samples of *S. Scripta* were collected from the Vellar estuary (11°29' N. Long; 79°46 E. Lat.), during January, 1988,

Table 1: Percentage of reduction in protein, carbophydrate and lipid at different concentration levels in whole body tissue and different body parts of *Sunetta scripta*

Sample	*Concentration*	*Protein*	*Carbohydrate*	*Lipid*
Wholy body tissue	Control	100.00	100.00	100.00
	10 ppb	87.93	87.88	83.33
	20 ppb	71.98	63.64	66.67
	30 ppb	61.64	54.55	62.50
	60 ppb	61.21	51.52	43.75
Mantle	Control	100.00	100.00	100.00
	10 ppb	97.87	98.55	96.00
	20 ppb	83.40	95.65	84.00
	30 ppb	28.40	94.20	30.00
	60 ppb	28..00	91.30	18.00
Gill	Control	100.00	100.00	100.00
	10 ppb	94.69	97.93	91.07
	20 ppb	87.35	91.03	75.00
	30 ppb	84.08	89.66	75.00
	60 ppb	80.82	88.28	69.64
Digestive diverticula	Control	100.00	100.00	100.00
	10 ppb	92.31	90.12	96.77
	20 ppb	84.62	90.12	90.32
	30 ppb	80.77	81.48	77.42
	60 ppb	69.83	67.90	67.42
Gonad	Control	100.00	100.00	100.00
	10 ppb	97.09	94.29	89.99
	20 ppb	91.91	91.43	82.86
	30 ppb	84.79	82.86	77.14
	60 ppb	77.99	68.57	68.57
Adductor muscle	Control	100.00	100.00	100.00
	10 ppb	96.94	98.08	93.33
	20 ppb	95.92	96.15	93.33
	30 ppb	91.84	82.69	86.67
	60 ppb	90.82	71.15	66.67

They were immediately transferred to the aquarium and allows them for a period of five days before experimentation to get acclimate to the laboratory conditions. They were fed daily. Similar size groups (34.5 to 35.0 mm) of animals selected and exposed to 10, 20, 30 and 60 ppb endosulfan. Acetone controls were also run along the exposure. The experiment was run for 96 hours. After 96 hours 5 animals from each concentration were taken out and washed gently with distilled water after removing the shells. Whole body and different body organs (mantle, gill, digestive diverticula, adductor muscle and gonad) were separated and ovendried for 24 hours at 40°C ±2°C. Appropriate amount of ground tissue samples utilized for evaluating protein content by Biurete method[12] carbohydrate by sulphuric acid – phenol method[2] and lipid by chloroform – Methanol method[3.]

RESULTS AND DISCUSSION

The level of protein in *S. scripta* were shown in Table 1. At the highest concentraiton of 60 ppb reduced nearly 38.79 percent of protein was observed from the whole body tissue. Whereas in different body parts maximum reduction in protein level was 30.07% in the digestive diverticula with minimum of 9.18% in the adductor muscle. The pattern of reduction in protein level observed as 38.79% in whole body tissues > 30.07% in digestive diverticula > 23.4% in mantle > 22.01 % in gonad > 19.18 % in gill > 9.18 % in adductor muscle.

The reduction in protein level not only concentration dependent but also vary with different body region. The reduction in protein at 10 ppb was 12.07 % from the whoe body tissue whereas in different body parts ranging from 2.13 % 7.69 % in the mantle and digestive diverticular respectively.

A similar trend was recorded for carbohydrate as that of protein content of *S. scripta* exposed to the pesticide. At 60 ppb reduction in carbohydrate content in whole body was 48.48 %, 8.7 % in the mantle and 32.1% in the digestive diverticula.

The lipid content of *S. scripta* was observed in both treated and control. From the untreated animal containing 11.65% of lipid (whole body tissue) reduction in lipid level at 60 ppb was 56.25% in the whole body. At lower concentration (10 ppb) reduction in lipid was 10.01% in the gonald and 3.23 % in the digestive diverticula. At higher (60 ppb) concentration reduction in lipid was ranged from 31% to 82% in gill and matnle respectively.

The bivalve constitute one of the major sources of cheap nutrition next to fishes and prawns for the human beings[6.] The nutritional value of different lamellibranchs on their various nutrient reserves. Since changes in chemical composition induced in different tissues of molluscs by pollutant do not seem to have been studied an signigicant detail [8, 11]. The decline on the protein suggests an intensive proteolysis in respective tissues.

The present results show that endosulfan reduce the chemical composition in all tested tissues showed a significant lipid decline. Decrease in the lipid content in the present investigation might be due to utilization of lipids for the energy demand associated with the situation of stress.[5]

REFERENCES

1. Dravidamani, M., 1985, Studies on the biology of *Sunetta scripta* (Mollusca : Bivalvia : Veneridae) from Portonovo coast. M. Phil. Thesis, Annamalai University.
2. Dubois, M., K.A. Giller, J.K. Hamilton, R.A. Reboxs and F. Smith, 1956, Calorimetric method for determination of sugars and related substances. Analyt. Chem., 28: 350-356.
3. Folch, J., M. Lees and G.G. Sloane Stanby, 1956, A simple method for the lipid isolation and purification of total lipids from animal tissues. J. Biol. Chem., 2256; 407-509.
4. GESAMP, 1976, IMCO/FAO/UNESCO/WHO/IAEA/UN/Joint Group of Experts on the Scientific Aspects of Marine Pollution. Review of harmful substances. Rep. Stud. GESAMP, 2:80.
5. Harpert, 1977, A Review of Physiological Chemistry 16 Edn. California Lange, Medical Pub., p. 269.
6. Jeyabal, R., 1986, Studies on *Meretrix meretrix* (Mollusca: Bivalvia). Ph.D. Thesis, Annamalai University.
7. Johnson, W.W. and Foster, L. Mayer, 1972, Pesticides and the aquatic environment.
8. Kulkarni, A. B. and V. N. Utkar, 1983, Effect of molluscicide on the biochemical composition of the tissue in the fresh water prosobranch snail, *Viviparus hengalensis*, J, Environ Biol., 4(4) : 215-220.
9. Palanichamy, S., P. Baskaran and M.P. Balasubramanian, 1986, Sublethal effects of selected pesticide on protein, carbohydrate and lipid content of different tissues of *oreochromis mossambicus*. Proc. Pest. Resid. and Env. Pollu., pp. 97-102.
10. Ramalingam, K., 1980, Studies on the effects of sublethal concentraiton of few

toxicants on biochemistry, physiology and histology of *Tilapia mossambica* (Peteres). Ph.D. Thesis, University of Madras.

11. Rao, K.R., P.A. Kulkarni, K.S. Pillar and U.H. Mane, 1987, Effect of fluoride on the fresh water bivalve mollusc, *Indonaia caerulus* on relation to effect of pH, accumulation and detoxification and respiration. Proc. Nat. Symp. Ecotoxic., pp. 13-23.

12. Raymont, J.E.G., A. Austin and E. Lingford, 1964, Biochemical studies on marine zooplankton. Biochemical composition of *Neomysis integer.* J. Cons. Perm. Explor. Mer., 28: 354-363.

13. Subba Rao, G., 1987, Biologoical effects of pesticides on the edible crab *Scylla serrata*. III. Carbohydrate metabolism and survival at sub zero temperature. Ph. D. Thesis, Annamalai University, p. 160.

14. Umminger, B. L., 1970, Physiological studies on supper cooled hill fish *Pundulus heteroclitas*. III. Carbohydrate metabolism and survival at sub zero temperature. J. Exp. Zoo., 173 : 159-174.

HEAVY METALS CONCENTRATION IN THE SEDIMENTS FROM THE SOUTH EAST COAST OF INDIA

C. STELLA, T. SARADHA, J. TAGORE and P. KASHINATHAN,
Centre of Advanced Study in Marine Biology,
Annamalai University, Parangipettai,
Tamil Nadu 608 502.

ABSTRACT

Concentration of Copper, iron, zinc and manganese were monitored for 6 months (July – Dec. 1987) in sediments taken from 5 different beaches of South East Coast of India. In all areas, marked variations on heavy metals concentration were observed. Sediments of Cuddalore Coast showed comparatively high concentration of heavy metals than other regions.

INTRODUCTION

Heavy metals are discharged into estuaries and Coastal waters with domestic and industrial wastes are present as particulates and it has strong affinities for solid phases in the receiving water.[6] But they are relatively insoluble in water. The concentration of heavy metals in sediments have been used for detecting comparative environmental data or regional differences in the dissolved heavy metal content of estuaries and coastal waters. Commonly, aquatic organism especially filter feeders have been favoured in these monitoring surveys of heavy metals. But in the present study, sediment samples were used to monitor the dispersal and density of heavy metal loads in these areas.

MATERIAL AND METHODS

The sediment samples were collected from Cuddalore, Karaikal, Nagapattinam, Velankanni and Tranquebar intertidal areas from July to December (1987) at montly intervals. They were collected by vertical gravity corer. The samples were labelled and dried in an oven at 100°C for 24 hrs. The heavy metals were determined using the combined and reducing technique[2] The reduced samples were analysed using Atomic Absorption Spectrophotometer (Techtron AAS, Model 1100) with air acetylene flame.

RESULTS AND DISCUSSION

The concentration of heavy metals in the present study area were presented in Tables 1-4. The sediment concentration of iron and manganese was comparatively high in Cuddalore intertidal area than the other coasts. (Fe 800 1025µg/g and Mn 225–

Table 1: Concentration of dissolved iron in sediments (μg/g)

Months	*Cuddalore*	*Karaikal*	*Nagapattinam*	*Velakanni*	*Tranqueber*
July	600	525	500	425	625
August	875	400	400	300	500
September	925	600	625	450	575
October	900	625	700	425	600
November	1025	750	750	600	650
October	1000	800	600	575	700

Table 2: Concentration of dissolved manganese in sediments (μg/g)

Months	*Cuddalore*	*Karaikal*	*Nagapattinam*	*Velankanni*	*Tranqueber*
July	250	275	250	100	150
August	225	300	175	150	100
September	350	200	250	200	300
October	300	275	400	275	275
November	375	325	375	350	325
December	375	300	300	300	300

Table 3 : Concentration of dissolved copper in sediments (μg/g)

Months	*Cuddalore*	*Karaikal*	*Nagapattinam*	*Velankanni*	*Tranque*
July	450	525	375	425	500
August	325	325	325	550	425
September	325	325	325	425	425
October	400	600	400	600	700
November	475	675	425	750	600
December	400	700	400	700	675

Table 4: Concentration of dissolved zinc in sediments (μg/g)

Months	*Cuddalore*	*Karaikal*	*Nagapattinam*	*Velankanni*	*Tranqueb*
July	400	425	375	300	400
August	475	400	425	300	625
September	325	400	600	425	500
October	500	475	625	375	575
November	525	500	725	600	625
December	500	525	700	625	800

375μg/g). But the concentration of copper and zinc in Cuddalore coast were lower than the other coast.

The concentration of iron in sediment ranged from 300 to 1025 μg/g and in the other coast iron concentration varied from 300 to 800 μg/g. During November, the maximum concentration of iron was recorded in Cuddalore, Nagapattinam and Velankanni and in December it was recorded in Karaikal and Tanqueber. But the minimum concentration of iron was recorded during July and August (Table 1).

In sediment, the concentration of manganese ranged from 100 to 400 μg/g. Manganese concentration in Cuddalore coasat wsa recorded from 225 to 375 μg/g and in other coasts these concentration ranged from 100 to 400 μg/g. During November, the maximum concentration of Mn was recorded in Cuddalore, Karaikal, Vellankanni and Tranqueber and in October if was recorded in Nagapattinam. The minimum concentration of manganese was observed during July and August (Table 2).

The concentration of zinc in sediment ranged from 300 to 800 μg/g. InTranquebar coast the high concentration of zinc ranged from 400 to 800 μg/g and in other coasts these concentration varied from 300 to 725 μg/g. In these cases the maximum concentration of zinc was recorded during November and December and the minimum concentration was recorded in July and August (Table 4).

In the present study, the heavy metals concentration were observed at different beaches of South east coast of India. The investigation reveals that the heavy metals in the sediment showed high concentration which occurs during November and December and the low concentration were observed in July and August.

The higher concentration of heavy metals in the sediments were influenced by arthopogenic introduction and land drainage. Pragadeeswaran *et al.*[8] described the Kodaikkari sediment concentration. It receives a large amount of river runoff and agricultural and other land drainage. It may also be responsible for the increased level of metals in sediment. Higher concentration of iron and manganese in Cuddalore coast was also influenced by higher organic content, whereas in the present study area the organic content was recorded to be less than 4%.

Pettijohn[7] is of the opinion that the factors such as velocity, diretion and stability of the transporting currents, the depth of the above water sediment and the turbulence of the medium has an important bearing on sedimentation of the particles. Similar influence of environmental parameters and the removal of metals from the sediment has also been reported by Bryan[1], Mahendran[5] and Senthilnathan[9]. The sediment serves as an important reservoir for most heavy metals[10]. According to Drniker and Notting[3], the metal concentration in the sediment is more in fresh water zone than in marine zone.

Seasonal variations in the metal concentration were also noticed in the present study. Maximum values of metals were observed during November and December and the minimum were recorded in July and August. Kumaraguru[4] and Pragadeswaran[8] reported that the high metal concentration was recorded during monsoon.

ACKNOWLEDGEMENT

The authors wish to express their sincere thanks to the Director, CAS in Marine Biology, for providing all facilities and thankful to ICAR, UGC, New Delhi, for the financial support.

REFERENCES

1. Bryan, C. W., 1976, Heavy metal contamination in the sea. In: Marine pollution (Ed.) Johnston, R. Academic Press, London, New York, pp. 185-302.
2. Chester, R. and M. J. Hughes, 1967, A chemical technique for the separation of ferro manganese minerals carbonate minerals and adsorbed trace elements from pelagic sediments. Chem. Geol., 2: 249-262.
3. Drniker, J.C. and R.F. Notting, 1976, Neth, J. Sea Res. 10, 71.
4. Kumaraguru, A.K., 1980, Studies on the chemical and biological transport of heavy metal pollutants copper, zinc and mercury in Vellar estuary and the toxicity of three pollutants to some estuarine fish and shellfish, Ph.D. Thesis, Annamalai University, Portonovo, India.
5. Magendran, A., 1985, Cycling of zinc in the Vellar estuary, M. Phil. Thesis, Annamalai University.
6. Moral, F.M.M., J.C. Westall, C.R.O.'Melia, and J. J. Morgan, 1975, Fate of trace metals in Los Angels Country waste water discharge, Environ Sci Technol., 9: 756-761.
7. Pettijohn, F. J., 1957, Sedimentary rocks II edition New York. Harper and Row.
8. Pragadeeswaran, V., P. Anhazhagan, E.Karthikeyan and R.Natarajan, 1987, Distribution of iron and manganese in the Kodiakkarai swamp, Palk Strait and Bay of Bengal. Electro Chem., 3(5): 375-377.
9. Senthilnathan, S., 1985, Cycling of iron in the Vellar estuary. M.Phil. Thesis, Annamalai University.
10. Wolfe, D.A., 1975, Modelling the distribution and cycling of metallic elements in estuarine ecosystem. In Estuarine Research, Vol. I : Chemistry Biology and the estuarine system. E. Croin (Ed.), Academic Press, pp. 645-671.

CONCENTRATIONS OF COPPER AND ZINC IN THE TISSUES OF *SANGUINOLARIA DIPHOS* (GMELIN) (MOLLUSCA: BIVALVIA)

J. Tagore, T. Saradha and R. Kasinathan
Centre of Advanced Study in Marine Biology,
Protonov-608-502,
Tamil Nadu, India.

ABSTRACT

The concentration of trace metals in the *Sanguinolaria diphos* of the mouth of Vellar estuary is observed for the period of one year (January to December, 1987). Copper and zinc contents of the different body parts such as gill, mantle, labial palp, siphon, adductor muscle, foot, gonad and stomach were estimated. The result was compared with the concentration of metals present in the habitat of the organism Viz., water and sediment. The distribution of metal- load in the tissues exhibited seasonality.

Introduction

Apart from the may analyses of sea water and the nutritional studies which have indicated that filter feeding bivalves the trace elements to considerable extent. Now the use of inorganic salts rich in metals along with manures and fertilizers in becoming a general practice among the farmers as a supplement of mineral requirement of the plant or mineral deficient soils. The copper and zinc are coming under the essential heavy metals to the life of marine organisms. The zinc in their optimum concentrations are essential for the normal proper growth and it becomes toxic when it exceeds the optimum level. Molluscs have been favoured for monitoring surveys of metallic pollution, since they accumulate higher concentration of metals than other group. Sankarn *et al*[7] reported the concentration of coper and zinc in *Crassostrea madrasensis* from the Cochin region. Kumaraguru[3] described the *chemical* and biological transport of heavy metal pollutants like Cu, Zn and Hg and their toxicity to some estuarine shell fishes. Lakshmana and Nambisan studied the seasonal variation in trace metal content in three bivalves Viz., *villorita cyprinoides* var. *Cochinensis. Meretrix casta* and *Perna viridis.* Magendran[4] and Senthilnathan have done a detailed work on the cycling of trace metals in the Vellar estuary. Tagore *et al.* 9.10.11 have investigated the trace metals content in estuarine bivalves. Premanathan *et al* made a monitoring surveys in lamellibranchs and Rajendran *et al.* studied the trace matal content in Cuddalore Oysters.

Since *Sanguinolaria diphos* is found in the mouth of the Vellar estuary which is influenced by the river flow, man made pollution and other hydrographical factors, any fluctuation that occurs in the quality of water is reflected in the quality of the animal tissue. In this present study, the lamellibranch *Sanguinolaria diphos* are used to monitor the dispersal and density of metal loads in the Vellar estuary. Hence the study was made to know the levels of copper and zinc in the different soft tissues such as gill, mantle, labial palps, foot, siphon, adductor muscle, stomach and gonad of the *S. diphos* and the results are compared with its habitat Viz., water and sediment.

Material and Methods

Specimens of *S. diphos* were collected at monthly intervals for a period of one year (January to December, 1985) from the clayed sediment of mouth of Vellar estuary near Portonovo coast (11^{0}29'N Long; 79^{0}46'E Lat.). Animals ranging from the same size groups (30 cm -31 cm.) were taken from this study. The collected specimens were kept in filtered sea water for one day to clear the stomach contents. The gill, mantle, labial palps, siphone., foot adductor muscle stomach, crystalline style and gonad were carefully isolated, washed with distilled water and dried in an oven 110^{0}C to one day. Tissues were always kept out of contact with metal instruments or particles. These tissues were digested with perchloric acid and nitric acid in the ratio of 1:3[12] until a clear solution was obtained. The digested sample was then made upto 25 ml with 5 nitric acid and aspirated into AAS.

Results and Discussion

The concentration of Zn was relatively more than that of Cu in all the body parts analysed (Table 1 and 2). Seasonal variations in the heavy metal concentrations were also discernible from the data. High values of Zn and Cu were observed during monsooon (October) and early postmonsoon (January) and low at summer. The trace metal accumulation in different soft body tissues, showed that gill and mantle accumulated more than the other tissues. The distribution of metal load was in the following decrease ordersediment, followed *S. diphos* and water.

Seasonality in heavy metal levels has been reported by Senthilnathan *et al.* High Zn and Cu levels in *C. madrasensis* has Cu and Zn were found to be accumulated more in the gill and mantle of *s.diphos.* Similar observations have been reported earlier in

Months	Gill		Mantle		Labial Palps		Foot		Gonad		Siphon		Adductor muscle		crystalling style		Stomach	
	Cu	Zn	Cu	Zn	Cu	Zn	Cu	Zn	Cu	Zn	Cu	Zn	Cu	Zn	Cu	Zn	Cu	Zn
January	4.0	39.0	2.6	44.2	19.0	99.0	15.0	36.9	7.8	39.2	4.0	56.9	4.0	29.0	3.8	26.8	6.8	19.8
Fabruary	3.0	36.0	2.9	39.0	16.9	86.5	14.2	31.1	6.9	41.0	3.9	26.8	2.9	28.0	6.0	28.0	6.0	17.8
March	2.0	33.2	3.1	36.8	15.0	77.1	11.9	26.9	7.1	36.8	4.1	52.8	3.5	28.0	1.1.	22.8	5.8	18.0
April	1.0	31.0	3.5	29.9	14.2	90.0	10.1	24.2	7.0	35.9	4.0	50.0	3.0	20.0	1.0	21.0	4.5	18.8
May	5.0	44.1	3.9	59.0	21.0	100.0	18.1	42.0	9.0	41.0	5.0	60.1	4.1	38.0	4.8	32.0	7.5	22.8
June	3.1	26.9	6.7	6.2	48.0	99.0	32.5	15.9	19.9	14.4	10.1	120.0	9.8	79.0	9.0	61.8	9.0	50.0
July	3.2	55.0	2.1	6.1	11.0	107.0	11.2	15.0	45.15	150.0	4.9	89.0	3.5	30.0	4.0	27.0	6.0	24.0
August	2.9	26.0	8.8	5.9	45.0	92.5	32.0	15.8	18.1	132.0	9.2	111.0	8.8	72.0	9.0	52.0	8.5	48.0
September	3.0	25.1	3.9	5.4	44.2	583.7	32.0	158.0	17.3	128.0	8.5	118.0	8.8	77.0	8.0	89.0	8.0	48.0
October	4.5	27.5	4.8	67.0	33.7	244.7	23.25	87.8	18.2	86.75	9.5	119.5	818	59.0	88.0	7.8	7.8	68.9
November	2.3	143.7	1.9	58.0	56.0	185.0	17.8	301.7	11.5	300.0	10.5	88.8	108	112.0	29.0	10.8	7.8	14.8
December	1.3	99.1	0.5	26.8	85.7	347.7	65.7	279.0	56.0	265.0	8.0	51.5	8.8	74.8	195.0	11.0	7.8	90.0

Table 2 : Bio-concentration of heavy Metals Zn & Cu in sediment and water

Months	*Sediment*		*Water*	
	Zinc	*Copper*	*Zinc*	*Copper*
January	81.90	9.10	80.50	10.75
February	85.60	8.90	78.00	9.50
March	84.70	8.50	54.50	9.25
April	82.90	9.00	53.00	80.21
May	45.90	10.10	52.00	8.97
June	35.80	9.80	49.50	9.75
July	99.90	20.01	44.25	12.00
August	98.10	19.00	43.02	13.00
September	98.50	19.25	41.10	12.50
October	18.90	7.75	83.75	11.50
November	125.25	12.25	128.00	10.25
December	151.25	7.75	114.50	9.50

C. madrasensis. As gills are responsible for the water flow in bivalves they are continuously exposed to metal accumulation than the other body parts. The mucous secretion covering the gills and mantles also cause high metal concentration as reported earlier in *C. madrasesis, Ostrea edulis*[1] *and C. virginica*[13]. The filter feeding nature of bivales is also a reason for the high accumulation of metals in the gills. Influx of high amount of freshwater due to monsoonal rains may be causative factor for the increase in the heavy metal concentration and subsequent bio-accumulation in the clams monsoon and postmonsoon seasons. The Zn and Cu concentration was found to be more in sediment than all soft tissues of the *S. diphos.*

REFERENCES

1. Gorge, S.G., B.J.S. Pirie, A.R. Cheyne, T.L. Coombs and P.T. Grant, 1978, *Mar. Biol.*, 45:147.
2. Lakshmanan, P.T. and P.N.K. Nambisan, 1983, *Indian J. Mar Sci.* 12:100.
3. Kumaraguru, A.K., 1980, Studies on the chemical and biologocal transport of heavy metal pollutants copper, zinc and mercury in Vellar estuary and the toxicity of these pollutants to some estuarine fish and shall fish, Ph. D. Thesis, Annamalai University.
4. Magendran, A., 1985, Cycling of zinc in the Vellar estuary. M. Phill Thesis. Annamalai University.
5. Premanathan, M.J., Tagore, R. Kasinathan and K. Kathiresan, Variation on heavy metal concentration in bivalves collected from Portonovo waters. Proc. Nat. Symp. Ecotoxic. pp. 129-133. 1987.
6. Rajendran, N., Tagore and R. Kasinathan, 1988, Heavy metal concentration in oyster *Crassostrea madrasensis* (Preston) of Cuddalore back waters. *Indian J. Mar. Sci.* (In press).
7. Sankaranaryan, V., K.S. Purushan and T.S.S. Rao, 1978, *Indian J. Mar. Sci.*, 7: 130.
8. Senthilnatha, S., 1985, Cycling of Iron in the Vellar estuary. M. Phill Thesis, Annamalai University.
9. Tagore, J., A. Shanmugam and R. Kasinathan, 1987a, Bio-accumulation of iron and manganese in *Sunetta scripta* (Linnaeus) from Portonovo coast. *Proc. Nat. Symp. Ecotoxic.* pp. 125-128.

10. Tagore. J., A. Shanmugam, T. Saradha and R. Kasinathan, 1987b, Metalic concentration in the gonads of bivales from Portonovo waters. Paper presented in the Vth All Indian Symposium on Invertebrate Reproduction. Palani, 24-27th Sept. 1987.

11. Tagore. J., R. Kasinathan, V.S. Chandrasekaran and A. Shanmugam, 1988, Seasonal Variations of the heavy metals Cu and Zn in different body part of a venerid clam *Sunetta scripta*. *Indian J. Mar. Sci.*, (In Press).

12. Topping, G., 1973, Heavy metals in shall fish from Scottish waters. Aquaculture. 1 (4) 379-384.

13. Wolf, D.A., 1975, Modelling the distribution and cycling of metallic elements in estuarine ecosystem. In estuarine Research, Vol. I : Chemistry, Biology and the estuarine system E. Cronin (Ed.) Academic Press, pp. 645-671.

BIBLIOGRAPHY

Abel, P.D., 1974. Toxicity of synthetic detergents to fish and aquatic invertebrates. J. Fish. Biol. 6 : 279-298.

Abraham, K.C., 1953. Observations on the biology of *Meretrix casta* (chemistry) J. Zool. Soc. India, 5 : 163-190.

Abraham, T.J., K.Y.M. Salih and J. Chacko, 1986. Effects of heavy metals on the filtration rate of bivalve *Villorita cyprinoides* (Hanley). India J. Mar. Sci., 15 : 195-196.

Achuthankutty, C.T., A. Stirling, S. Nair, P.A. Loka Bharathi and M.R. Menezes, 1978. Sandy beach at Baine, Goa; its ecology and production. Indian J. Mar. Sci. 7(7) : 23-29.

Ajmalkhan, S. and R. Natarajan, 1981. Salinity tolerance of hermit crabs. Indian J. Mar. Sci., 10 : 393-395.

Alagarswami, K., 1966. Studies on some aspects of biology of the wedge clam *Donax faba* (Gmelin) from Mandapam coast in the Gulf of Mannar. J. Mar. Biol. Ass. India, 8(1) : 56-75.

Allen, J.A., 1958. The basic form and adaptations to habitat in the Lucinacea (Eulamellibranchia) Phil. Trans. R. Soc. London, B 241 : 421-484.

Anderson, 1972. The effects of oil on the gill filtration rate of *Mya arenaria.* In Va. J. Sci., (23) 457 pp.

Anonymous, 1951. Soil survey manual. U.S. Department Agriculture hand book. No. 18. p. 210-211.

Ansell, A.D., 1961. The functional morphology of the species of Venericea (Eulamellibranchia). J. mar. biol. Ass. U.K., 41 : 489-515.

Ansell, A.D., 1972. Distribution, growth and seasonal changes in biochemical composition for the bivalve *Donax vittatus* (da costa) from Kames Bay, Millport, J. exp. mar. Biol. Ecol., 10 : 137-150.

Ansell, A.D., 1974. Seasonal changes in biochemical composition of the bivalve. *Nucula sulcata* from the Glyok Sea are. Mar. Biol., 25 : 101-108.

Ansell, A.D., 1974. Seasonal changes in biochemical composition of the bivalve *Abra alba* from the Clyde sea area. Mar. Biol., 23 : 13-20.

Ansell, A.D., Barnett, P.R.O., Bodoy, A. and Masse, H. 1980. Upper temperature tolerances of some European Molluscs. I. *Tellina fabula* and *T. tenuis.* mar. Biol., 58 : 33-39.

Ansell, A.D. and Bodoy, 1979. Comparison of events in the seasonal cycle for *Donax vittatus* and *D. truncutus* in European waters. Proc. 13th Europ. Mar. Biol. Symp. 191-198.

Ansell, A.D. and K.F. Lander, 1967. Studies on the hard shell clam, *Venus mercenaria.*

Ansell, A.D., F.A. Loosmore and K.F. Lander, 1964. Studies on the hard shell clam, *Venus mercenaria* in condition and biochmeical composition. *J. appl. Ecol.*, 1 : 83-85.

Ansell, A.D. and A. Trevllion, 1967. Studies on *Tellina tenuis* da costa. I. seasonal growth and biochemical cycle. J. exp. mar. biol. Ecol., 1 : 220-235.

Atkins, D., 1937. On the ciliary mechanisms and interrelationships of the Lamellibranchiata. Part I. Some observations on the ciliary feeding mechanism of *Glycymeris glycymeris* (L) and *Arca tetragona* (Pol). Quart. J. Microscop. Sci., 79 : 181-308.

Arnold, D.C., 1972. Salinity tolerance of some common prosobranchs. J. Mar. Biol. Ass. U.K., 58 (2) : 476-486.

Bachelet, G., 1980. Growth and Recruitment of the Tellinid Bivalve *Macoma balthica* at the Southern Limit of Its Geogrophical Distribution, the Gironde Estuary (SW France) Mar. Biol., 59 : 105-117.

*Badman, D.G., 1975. Comp. Biochem. Physiol. 51(4): 741.

Balasubrahmanyan, K., 1964. Studies in the Ecology of the Vellar estuary. III. The intertidal and estuarine polychaeta. J. Annamalai Univ., 25 : p. 101-104.

Balasubramanyan, K., 1984. Studies on *Meretrix casta* (Chemnitz) (Mollusca: Bivalva:

Veneridae) from the Vellar estuary. Porto Novo. Ph.D. Thesis, Annamalai University.

Bloom, S.A., J.L. Simon and V.D. Hunter, 1972. Animal sediment relations and community analysis of a Florida Estuary. Mar. Biol., 13 : 43.

Boss, K.J., 1966. The sub family Tellininae in the Western Atlantic. The gerea *Tellina* (Part II) and Tellidora Johnsonia. 4(46) : 273-344.

O'Brein, R.D. Insecticides, Action and metabolism (Academic press, New York), 319.

Bridges, W.R., B.J. Kallman and A.K. Andrews, 1963. Persistence of DDT and its metabolites in a farm pond. Trans. Am. Fish. Soc., 92 : 421-427.

Broach, V., 1979. Habitat selection of two congeneric bivalves, *Cardium edule* (L) and *Cardium glaucum* (Bruguiere) in sympatric and allopatric populations. Mar. Biol., 54 : 149-156.

Brodtmann, N.V., 1970. Studies on the assimilation of 1, 1, 1-trichloro -2, 2 - bis (P-chlorophenyl) ethane (DDT) by *Crassostrea virginica*. Genelin. Bull. Environ. Contam. Toxicol., 5 : 455-462.

Brousseau, D.J., 1978. Spawning cycle, fecundity and recruitment in a population of soft shell clam *Mya arenaria* from Cape Ann. Massachusetts. Fish. Bull., 76(1) : 155-166.

Brown, B.E. and Newell, R.C., 1972. The effect of copper and zinc on the metabolism of the mussel *Mytilus edulis,* Mar. Biol., 16 : 108-118.

Butler, P.A., 1969. The significance of DDT residues in estuarine fauna. In chemical fallout Ed. by M.W. Miller and G.G. Berg. 205-220. Springfield, Illinois, Charles C. Thomas.

Butler, J.A., Milleman, R.E. and Stewart, N.E. 1968. Effects of the insecticides sevin on survival and growth of the cockle clam *Clinocardium nuttalli.* J. Fish. Res. Bd. Can., 25, 1621-1635.

Butler, P.A., Wilson, A.J., Jr., and Rick, A.J., 1962, Effects of pesticides on oysters. Proc. Nat. Shellfish. Ass., 51 : 23-32.

Cole, H.A. and Hepper, B.T. 1954. Use of neutral red solution for the comparative study of filtration rates of lamellibranchs. J. Cons., 20 : 197-203.

Cope, O.B., 1965. In research in pesticides, ed. Chichester Co. (Academic Press, New York) 115.

Davis, H.C., 1961. Effects of some pesticides on eggs and larvae of oysters. *(Crassostrea virginica)* and Clams *(Venus mercenaria)* Com. Fish Rev. 23 (12), 8-23.

Davis, H.C., Loosanoff, V.L. and Mackenzie, C.L., 1961. Field tests of a chemical method for the control of marine gastropods. Presented at the convention of the National shell fisheries Association, Baltimore, Maryland, August 1961. Bur. Commer. Fish. Biol. Lab., Milford. Cann. Bull., 25(3).

Dodgson, R.W., 1928. Report on mussel purification. Fish. Invest., London. (2) 10 (1) : 1-498.

Dinamanai, P., 1957. On the stomach and associated structures in the backwater clam *Villorita cyprinoides* (Gray) Var. *Cochinensis* (Hanley). Bull. Centr. Res. Inst. Univ. Kerala, 5 : 123-148.

Dinamani, P., 1967. Variatin in the stomach structure of the bivalvia. *Malacologia* 5 : 225-263.

Dravidamani, M., 1985. Studies on *Sunetta scripta* (Linnaeus) (Mollusca: Bivalvia; Veneridae) from Porto Novo Coast. M.Phil. Thesis, Annamalai University. 1985. p. 61.

Dubois, M., K.A. Giller, J.K. Hamilton, R.A. Robors and F. Smith, 1956. Colorimetric method for determinatin of sugars and related substances. Analyt. Chem., 28 : 350-356.

Dunning, A. and C.W. Major, 1974. The effects of cold seawater extracts of oil fractions upon the blue mussel, *Mytilus edulis*. In: Pollution and physiology of marine organisms (ed. F.J. Vernberg and W.B. Vernberg) pp. 349-366. New York, Academic Press.

Durve, V.S., 1964. Preliminary observations on the seasonal gonadal changes and spawning in the clam *Meretrix casta* (Chemnitz) from the Marine fish farm. J. mar. biol. Ass. India, 6(2) : 241-248.

Dwivedi, S.N., R.M.S. Bhargava, A.H. Parulekar, R.A. Selvakumar, S.Y.S. Singbal and V.N. Sankaranarayanan, 1973. Ecology and environmental monitoring of Mandovi, Zuari and Cumbarjua canal complex during monsoon months. J. Indian Fish. Assoc., 3 : 113-130.

Dye, A.H., 1977. An ecophysiological study of the benthic meiofauna of the Swartkps estuary. M.Sc., thesis, University Port Elizabeth.

Dyer, K.R. and K. Ramamoorthi, 1969. Salinity and water circulation in the Vellar estuary. Limnol. Oceanogr., 14 : 4-15.

Edmondosn, C.H., 1920. The reformation of the crystalline style in *Mya arenaria* after *extraction.* J. Exptl. Zool., 30 : 259-291.

Edwards, R.R.C., 1978. Effects of water soluble fraction on metabolism, growth and carbon budget of the shrimp *Crangon crangon.* Mar. Biol., 46 : 259-265.

Eisler, R., 1977. Acute toxicities of selected heavy metals to the softshell clam, *Mya arenaria.* Bull. Environ. Contam. Toxicol. 17 : 137-145

Eisler, R., and G.R. Gardner, 1973. Acute toxicology to an estuarine teleost of mixtures of cadmium, copper and zinc salts. J. Fish. Biol., 5 : 131-142.

Eller, L.L., 1971. Histopathologic lesions in cut throat trout *(Salmo clarki)* exposed chronically to the insecticide endrin. Amer. J. Pathol., 64 : 321.

Emmanuel Rajan, J.T., 1985. The toxic effects of WSF of Diesel on *Anadara rhombea* (Born) M.Phil. Thesis, Annamalai University.

Engel, R.H. Neat, M.J and Hillman, H.J., 1972. Sublethal chronic effects of DDT and Lindane on glycolytic and glucogenic enzymes of the quahog, *Mercenaria mercenaria.* In: Marine Pollution and Sea Life, pp. 257-259. London. Fishing News.

Ferguson, D.E., J.L. Lundke and G.G. Murphy, 1966. Dynamics of endrin uptake and release by resistant and susceptible strains of mosquito fish. Trans. Amer. Fish. Soc., 95 : 335-344.

Fingerman, M., 1982. Keynote address. Proc. Symp. Physiol. Respon. Anim. Pollut. 1-15.

Folch, J., M. Lees and G.H. Sloane Stanley, 1956. A simple method for the isolation and purification of total lipids from animal tissues. J. Biol. Chem., 226 : 497-509.

Fong, W.C., 1976. Uptake and retention of Kuwait crude oil and its effects on oxygen uptake by the soft-shell clam, *Mya grenaria.* J. Fish. Res. Bd. Can., 33 : 2774-2780.

Fraenkel, G.S., 1968. The heat resistance of intertidal snails of Bimini, Bahamas: ocean springs, Miss., and Woods Hole, Mass Physiol. Zool., 41 : 1-13.

Frederick, P. and Thunberg, 1977. Respiratory response of cunners to silver fish. Mar. Poll. Bulletin. 18(2) : 40-41.

Frere, M.H., 1971. Requisite sampling frequency for measuring nutrient and pesticide movement with runoff waters. J. Agri. Food Chem., 19 : 837-839.

Frost, K.R. and G.W. Ware, 1970. Pesticide drift from aerial and ground applications. Agri. Eng., 51 : 460-464.

Galtsoff, P.S., 1964. The American oyster *Crassostrea virginica* Gmelin. Fishery. Bull. Fish. Wilde. Serv. U.S. 64 : 1-40.

Galtsoff, P.S., H.F. Prytherch, R.O. Smith and K. Koehring, 1935. Effects of crude oil pollution on oysters in Louisiana waters. U.S. Bur. Fish. Bull., 18 : 143-210.

Gardner, G.R. and P.P. Yevich, 1970. Histological and hematological responses of an estuarine telest to cadmium. J. Fish. Res. Bd. Can. 27 : 2185-2196.

George John., 1980. Studies on *Anadara rhombea* (Born) (Mollusca: Bivalvia: Arcidae) from Porto Novo coastal waters. Ph.D. thesis, Annamalai University, 190 pp.

GESAMP, 1976. IMCO/FAO/UNESCO/WMO/IAEA/UN Joint Group of Experts on the Scientific Aspects of Marine Pollution (GESAMP) Review of harmful substances. Reports and studies. GESAMP NO. 2, 80 p.

Giese, A.G., 1959. Comparative physiology, annual reproductive cycles of marine invertebrates. Ann. Rev. Physiol., 21 : 547-576.

Giese, A.C., 1969. A new approach to the biochemical composition of the mollusc body. Oceanogr. Mar. Biol. Ann. Rev., 7 : 115-129.

Giese, A.C., M.A. Hart, A.M. Smith and M.A. Cheang, 1967. Seasonal changes in body component indices and chemical composition in Pismo clam. *Tivela stultorum.* Comp. Biochem. Physiol., 22 : 549-561.

Giese, A.C., A.N. Smith and M.A. Cheng, 1967. Seasonal changes in body component indices and chemical composition in Pismo clam *Tivela stultorum,* comp. Biochem. Physiol., 22 : 549-561.

Gillfillan, E.S., 1975. Decrease of net carbon flux in two species of mussels caused by extracts of crude oil. Mar. Biol., (Berle) 29 (1) : 53-58.

Gillfillan, E.S., D.W. Mayo, D.S. Page, D. Donovan and S. Hanson, 1977. Effects of varying concentrations of petroleum hydrocarbons in sediments on carbon flux in *Mya arenaria.* In: Physiological responses of marine biota to pollutants (ed. F.J. Vernberg, A. Calabrese, F.P. Thurberg and W.B. Vernberg), pp. 299-314. New York, Academic Press.

Gilfillan, E.S., L.C. Tiang, D. Donovan, S. Honson and D.W. Mayo, 1976. Reduction in carbon flux in *Mya arenaria* caused by a spill of No. 2. fuel oil. Mar. Biol., 37 : 115-123.

Gould, E. and J.J. Karolus, 1974. Physiological response of the cunner, *Tautoglabrus adspersus* to cadmium. V. observations on the biochemistry. In physiological response of the cunner, *Tautogolabrus adspersus* to cadmium. NOAA Tech. Rep. NMFS. SSRF - 681, 21-25.

Graham, A., 1934. The structure and relationship of lamellibranchs possessing a cruciform muscle. Proc. Roy. Soc. Edinb., 54 : 150-157.

Granham, A., 1949. The molluscan stomach. Trans. Ray. Soc. Edinburg., 61(3) 27 : 737-778.

Gravely, F.H., 1941. Shells and other animal remains found on the Madras beach, Part I - groups other than snails, etc. Bull. Madras Govt. Mus. Vol. V.

Gravely, 1942. Shells and other animal remains found on the Madras beach II, Snails Bull. Madras Govt. Mus., 62-63.

*Greeve, P.A. and S.L. Wit, 1971. Chem. Weegbl, 67, 7.

Hagerman, L., 1970. The oxygen consumption of *Crangon crangon* (Fabricius) (Crustacea, Natantia) ophelia, 7, 283-293.

Hagerman, L., 1976. The respiration during the moult cycle of *Crangon crangon* (Fabricius) (Crustacea, Natantia) ophelia, 15, 15-21.

Hanekom, N., 1975. A study of *Donax serra* in the estern cape. M.Sc., thesis, University Port Elizabeth.

Karkantra, S.N., 1975. Some observations on the clam beds of kali estuary, Karwar. Mahasagar Bull. NIO, 8 (land 2) : 101-108.

Hibbert, C.J., 1977. Growth and survivorship in a tidal-flat population of the Bivalve *Mercenaria mercenaria* from Southampton water. Mar. Biol., 44 : 71-76.

Holdon, A.V., 1962. A study of the absorption of 14C labelled DOT from water by fish. Ann. Appl. Biol., 50 : 467.

Holme, N.A., 1961. Notes on the mode of life of the Tellinidae (Lamellibranchiata), J. mar. biol. Ass. U.K., 41 : 699-703.

Homell,J., 1922. The common molluscs of South India, Madras fish. Bull., 14 : 97-210.

Honell,J., 1949. The study of Indian Molluscs. J, Bomb. N.JH. Soc., 48(3,4).

Hughes, R.N., 1970. An energy budget for a tidal flat population of the bivalve *Scorbicularia plana* (DaCosta) J. Anim. Ecol., 39 : 357-381.

Hughes, N., 1971. Reproduction of *Scrobicularia plana* (Pelecypoda: Semelide). The Veliger, 14(1) : 77-82.

Hyman, L.H., 1967. The Invertebrates. Vol. 6. Mollusca. McGraw Hill Book Company Inc., New York.

Jackim, E., J.M., Hamlin and S. Sonis, 1970. Effects of metal poisoning on five liver enzymes in the killifish *(Fundulus heteroclitus)*. J. Fish. Res. Bd. Can. 27, 383-390.

Jacob, J. and K. Rangarajan, 1959. Seasonal cycle of hydrographical events in Vellar estuary. Proc. First All India Cong. Zool., Part 2: 329-350.

Jagadeesan, G., 1974. Studies on the particulate matter of Vellar estuary and the nearshore waters of Proto Novo. Ph.D. thesis, Annamalai University, 256 pp.

Jambulingam, M., 1988. Sublethal effects of endosulfan and cadmium on histological orgasnization and Hexokinase activity in *Oreochromis* (Tilapia) *mossambicus* (Peters) M.Phil. thesis, Annamalai University.

Jayabal, R., 1984. Studies on Meretrix *meretrix* (Linnaeus) (Mollusca: Bivalvia: Veneridae) from Porto Novo waters. Ph.D. thesis, Annamalai University.

Judd, W., 1979. The secretions and fine structure of bivalve crystalline style sacs. Ophelia, 18 : 205-233.

Jutting, T. Van Banthem, 1943. Mollusca Lamellibranchia, Fauna Nederland, 12 : 1-477.

Kalyanasundaram, M., 1982. Studies on *Katalysia opima* from the Vellar estuary. M.Phil, Thesis, Annamalai University, 78 pp.

Kanakasabai, R., 1985. Studies on some Neritids (Archaeogastropoda: Mollusca) from the Vellar-Coleroon estuarine complex (South east coast of India) Ph.D. Thesis, Annamalai University, 228 pp.

Kasinathan, S., 1963. Effect of Insulin on the carbohydrate metabolism of *Meretrix casta* (chemistry) Proc. Ind. Acad. Sci., 58 (6) : 367-374.

Keck, R.T. and Don Maurer, 1975. A comparative study of the hand clam, gonad development cycle. Biol. Bull., 148 : 243-258.

Kennedy, V.S., 1977. Reproduction in *Mytilus edulis aoteanus* and *Aculacomy maoriana* (Mollusca: Bivalvia) from Taylors Mistake, New Zealand. N.Z.J. Mar. Freshw. Res., 11(2) : 255-267.

*Kinne, O., 1967. Physiology of estuarine organisms with special reference to salinity and temperature. General aspects. In: Estuaries. Ed. G.H. Lauff, Amer, Assoc, Advance. Sci. Publ., No 83, Washington, D.C., 525-540.

*Kinne, O., 1972. Marine Ecology, I. ed. 0. Kinne. Wiley. Interscince, New York - 821.

*Kinne, O., (ed), 1975. Marine Ecology. John Wiley, New York.

Kittredge, J.S., 1974. Bioassays indicative of some sublethal effects of oil pollution. In Proc. Marine. Tech. Soc. p. 891-897.

Klein, L., 1962. River pollution 2: Causes and effects. Butterworths, London.

Krishnamurthy, K., 1964. Studies on plankton and nutrients in the inshore and estuarine waters of Porto Novo, Ph.D. Thesis, Annamalai University, India, 239 pp.

Kristensen, J.H., 1972a. Structure and function of crystalline styles of bivalves. Ophelia, 19(1) : 91-108.

Kurian, C.V., R. Damodaran and A. Antony, 1975. Bottom fauna of the Vembanad lake. Bull. Dept. Mar. Sci. Univ. Cochin, Vii (4) : 987-994.

Lassig, J., 1965. The distribution of marine and brackish water lamellibranchs in the northern Baltic area. Comm. Biol. Soc. Sci. Fenn. 28(5) : 1-41.

Lee, R.F., R. Saueraeber and A.A. Betison, 1972. Petroleum hydrocarbon uptake and discharge by the marine mussel *Mytilus edulis* science (Wash. D.C), 177 : 344-346.

Lingaraja, T., 1980. Toxicity studies of pesticide on *Therapon jarbua* (Forskal) Ph.D. Thesis, Annamalai University, 170 pp.

Lingaraja, T., P. Sasibhushana Rao and V.K. Venugopalan, 1979. DDT induced ehtological changes in estuarine fish. Environ. Biol. Fish., 4 : 83-88.

Lingaraja, T., R.A. Selvakumar and V.K. Venugopalan, 1980. A simple continuously monitoring apparatus for respiration studies in Aquatic organisms. India J. Expt. Bio., 18 : 413-415.

Litchfield, J.T. and F. Wilcoxon, 1949. A simplified method for evaluating dose effect experiments. J. Pharmac. Exp. Therap. 96 : 99-113.

Loosanoff, V.L., 1947. Effects of DDT upon setting, growth and survival of oysters. Fishing Gaz, 64, 94-96.

Loosanoff, V.L. and H.C. Davis, 1950. Conditioning *Venus mercenaria* for spawning in winter and breeding its larvas in the laboratory. Biol. Bull., 93 : 60-65.

Lowe, J.I., Wilson, P.D., Rick, A.J. and Wilson, A.J. 1970. Chronic exposure of oysters to DDT, Toxaphene and parathion. Proc. Natn. Shellfish. Ass. 61, 71-79.

Lowe, J.I., Wilson, P.O., Rick, A.J., and WJlson, A.J., 1971. Chronic exposure of oyster to DDT, toxaphene and parathion.

Lund, E.J., 1967. Effect of Bleed waters, soluble fraction and crude oil on the oyster. In. Publ. Inst. Mar. Sci. niv. Tex (4) 38-41 pp.

MacInnes, J. and Thurberg, F., 1973. Effects of metals on the behavior and oxygen consumption of the mud snails. Mar. Pollut. Bull. 4, 185-186.

Maclnnes, J.R. and P.P. Thurberg, 1973a. Effects of metals on the behaviour and oxygen consumption of the mud snails. Mar. Pollut. Bull., 4(12), 185-186.

Mane, U.K., 1974. The growth and breeding habits of the clam *Katalysia opima* in kalbadevi estuary at Ratnagiri. Indian J. Fish., 21 : 386-398.

Mane, U..H., Kachole, M.S. and Pawar, S.S., 1979. Effects of pesticides and nacrotants on bivalve molluscs. Malacologia. 18 : 347-360.

Mane, U.H., Muley, D.V., and Akartes, S.R. 1983. Toxicity of Endosulfan to freshwater and estuarine bivalves. Indian J. Comp. Anim. Physiol., 1(2) 1-9.

Marimuthu, M., 1986. Studies on *Crossostrea madrasensis* Preston (Mollusca: Bivlvia: Ostreidae) from Cuddalore backwaters. M.Phil. Thesis, Annamalai University. 24 pp.

Mathur, D.S. and P.D. Rane, 1979. Histopathological changes in the liver and intestine of *Rana cyanophlyctis* (induced by endrin). Proc. Symp. Environ. Biol., 415-419 pp.

Mc Erlean, A.J., 1964. Characteristics of *Macoma balthica* populations in the middle Patuxent estuary. Chesaplake Sci., 5 : 200-208.

McIntyre, 1968. The meio-fauna and macrofauna of some tropical beaches. J. Zool. London, 156 : 377-392.

McLachlan, A. and N. Hanekom, 1979. Aspects of the biology, ecology and seasonal fluctuations in the biochemical composition of *Donex serra* in the East cape. S. Afr. Zool., 14(4) : 183-193.

McLusky, D. S., S. A. Nair, A Stirling and R. Bhargava, 1975. The ecology of a central west Indian beach with particular reference to *Donax incarnatus* Mar. Biol., 30: 267-276.

Miller, D. S. and Kinter, W. B. 1977. In: Physiological responses f Marine Biota to pollutants, eds. Vernberg F. J., Calabrase, A. Thurberg F. P. and Vernberg, W. B. (Academic Press, New York – San Fransico, London), 75.

Moore, H. B. and N. N. Lopez, 1970. A contribution to the ecology of the lamellibranch *Tellina alternata* Bull. mar. Sci., Gulf Caribb., 20 : 971-979.

Morton, B. S., 1969a. Studies on the biology of *Dreissena polymarpha* Part I. General anatomy and morphology. Proc. Malacol. Soc. London, 38 : 301-321.

Muley, D. V. and U. H. Mane, 1987. Malathion induced changes in oxygen consumption in two species of freshwater lamellibranch moluscs from Godavari River, Maharastra State, India. J. Environ. Biol., 8 : 267-275.

Murray, J. A. and L. M. Vaughan, 1969. Measuring pesticide drift at distances to four miles. Bull. Am. Meteorol. Soc., 50 : 6.

Nagabhushanam, R., 1956. The rate of water filtration in the Marine wood boring molluscs, *Martesia striata* Proc. Indian Acad. Sci., 43-223.

Nagabhushanam, R. and R.S. Deshmukh, 1974. Seasonal changes in body component indices and chemical composition in the estuarine clam, *Meretrix meretrix* (L). Indian J. Fish., 21 : 531-542.

Nagarajan, D., 1987. Effect of endosulfan on the physioloigcal activities of estuarine bivalve *Meretrix casta* (chemnitz) M.Phil. Thesis, Annamalai University.

Nair, A., 1978. Production ecology of a tropical sandy beach at Benaulm, Goa. Indian J. Mar. Sci., 7 : 163-167.

Narasimham, K.A., 1968. Studies on some aspects of biology and fishery of the cockle *Anadara granosa* (Linne) from Kakinada Bay. Proc. Symp. on Mollusc. Mar. Biol. Ass. India, 407-417.

Nayar, K., 1955. Studies on the growth of Wedge clam Donax cuneatus. Indian J. Fish., 2 : 325-348.

Neff, J.M., B.A. Cox, D. Dixit and J.W. Anderson, 1976. Accumulation and Release of Petroleum Drived Aromatic Hydrocarbons by four species of Marine Animals. Mar. Biol., 38 : 279-289.

Nelson, T.C., 1918. On the origin, nature and function of the crystalline style of lamellibranchs. J. Morphol., 31 : 53-111.

Nelson, T.C., 1925. Recent contribution to the knowledge of the crystalline style of lamellibranchs. Biol. Bull., 49 : 86-99.

Nelson-Smith, A., 1970. The problem of oil pollution of the sea Adv. Mar. Biol., 8 : 215-306.

Newman, M.W. and S.A. MacLean, 1974. Physiological response of the cunner, *Tautogolabrus adspersus* to cadmium, VI. Histopathology. In physiological response of the cunner, *Tautogolabrus adspersus* to cadmium, pp. 27-33. NOAA Tech. Rep. NMFS. SSRF 681.

Newell, G.E., 1964. Physiological aspects of the ecology of intertidal molluscs. Wilber and Yonge I. pp. 59-81.

Newell, R.C., 1973. Am. Zool., 13, 513.

Newell, R.C., 1979. Factors affecting the rate of oxygen uptake. In biology of intertidal animals. Ch. II. Marine Ecology Survey, Faversham, Kent.

*0dum, E.P., 1963. Ecology, Modern Biology Series. Holt Riecharl and Winston. Inc. New York. 150 pp.

Olivia, J., 1981. Ecological studies in the intertidal region of the Vellar estuary (Porto Novo, S. India), Ph.D. Thesis, Annamalai University.

Owen, G., 1955. Observations on the stomach and digestive diverticula of the Lamellibranchia. I. The Anisomyaria and Eulamellibranchia. Q.J. Microsc. Sci., 96 : 517-537.

Owen, G., 1959. The ligament and digestive system in the taxodant bivalves. Proc. Malac. Soc. London, 33 : 215-223.

Owen, G., 1966. Digestion. In: Physiology of Mollusca (Ed. K.M. Wilbur and C.M. Yonge) Academic Press, New York, 53-96 pp.

*Philips, D.J.H., 1976. Mar. Biol., 38(59).

Pohlo, R.H., 1967. Aspects of the biology of *Donax gouldi* and a note on evolution in Tellinacea (Bivalvia). The Veliger, 9(3) : 330-337.

Pragatheswaran, V., 1987. Effects on Zinc, Copper and Mercury on *Ambassis commersoni* Curier (Osteichthyes: Perciformes: Ambassidae) Ph.D. Thesis, Annamalai University.

Purchon, R.D., 1955b. The stomach in Protobranchia and septibranchia (Lamellibranchia). Proc. Zool. Soc. London, 127 : 511-525.

Purchon, R.D., 1957. The stomach in the filibranchia and pseudolamellibranchia. Proc. Zool. Soc. London, 129 : 27-60.

Purchon, R.D., 1958. The stomach in Eulamellibranchia. Stomach type IV. Proc. Zool. Soc. London, 131 : 487-523.

Purchon, R.D., 1960. The stomach in the Eulamellibranchia: stomach Type IV and V. Proc. Zool. Soc. London, 39 : 253-262.

Purchon, R.D., 1977. The biology of the Mollusca, Pergamon Press, Oxford.

Rai, H.S., 1932. The shell fisheries of Bombay Presidency. J. Bombay Nat. Hist. Soc., 35(4) : 826-847.

Rajendran, N., 1984. Organochlorine pesticide residues in Vellar estuary. Ph.D. Thesis, CAS in Marine Biology, Annamalai University, India.

Ramadhas, V., 1977. Studies on phytoplankton, nutrients and some trace elements in Porto Novo waters. Ph.D. Thesis, Annamalai University, India, 135 pp.

Ramalingam, R. and Y. Srinivasa Reddy, 1981. Acute histopathological effects of lindane (r-Benzene hexa chloride) on the liver of *Colisa lalia.* Curr. Sci., 59(13) : 578-580.

Rangarajan, K., 1958. Diurnal tidal cycle in Vellar estuary. J. Zool. Soc. India, 10 : 54-67.

Rao, K.S., 1967. Annual reproductive cycle of the Wedge clam *Donax cuneatus* L.J. mar. Biol. Ass. India, 9(1) : 141-146.

Rao, K.V., 1951. Studies on the growth of *Katelysia opima* Proc. Indo. Pacific. Fish. Council. Sec. II : 94-102.

Rao, P.S., 1980. Pesticides toxicity to fish eggs and juveniles. Ph.D. Thesis, GAS in Marine Biology, Annamalai University.

Ratnaker, A.V. and C. Aswathy, 1979. Toxicity of aldrin to *Cyprinus carpio*. Proc. Symp. Environ. Biol. 377-382 pp.

Raymont, J.E.G., A. Austin and E. Lingford, 1964. Biochemical studies on marine zooplankton. I. Biochemical composition of *Neomysis integer.* J. Cons. Pern. Explor. mer., 28 : 354-363T

Redding, J.M. and R.L. Cory, 1975. Macroscpic benthic fauna of three tidal creeks adjoining the Rhode River, Mary land. Water resources Investigations. 39-75, U.S. Geological Survey. 23 pp.

Rhoads, D.C., 1970. Mass properties, stability and ecology of marine muds related to burrowing activity. In: Trace fossils, Eds. T.P. Crimes and J.C. Harper Seel House Press, Liverpool, 391-406.

Rhoads, D.C., 1974. Organisms-sediment relations on the muddy sea floor. Oceanogr. Mar. Biol., Ann. Rev., 12 : 263-300.

Rhoads, D.C. and D.K. Young, 1970. The influence of deposit feeding organisms on sediment stability and community trophic structure. J. Mar. Res., 28 : 150-178.

Richard Ivell, 1979. The biology and ecology of a brackish Lagoon bivalve, *Cerasderma glaucum* Bruguiera in an English Lagoon, The Widewater, Sussex, J. Moll. Stud., 45 : 385-400.

Roberts, D., 1975. Sub-lethal effects of chlorinated hydrocarbons, 6(2).

Rogers, W.E., 1959. Gonad development and spawning of the soft clam, *Mya arenaria.* Marine and Tide water news, 15 : 9-10.

*Salanki, J., and Lukacsovic, S.F. 1967. Biological kutatointe Zet-Annales (Hungary), 34, 85.

Salih, M., 1973. On the growth of the backwater clam, *Meretrix casta* (chemnitz) in the clam beds of Cochin bar mouth. J. Mar. Biol. Ass. India, 15 (1) : 345-353.

Sanders, H.L., 1958. Benthic studies in Buzzards Bay. I. Animal sediment relatinships. Linnol. Oceanogr., 3 : 245-258.

Santharam, K.R., B. Thayumanavan and S. Krishnaswamy, 1976. Toxicity of some insecticides to *Daphnia carinata* king an important link in the foodchain in the freshwater ecosystems, Indian J. Ecol., 3 : 70-73.

Saraswathy, M. and N. Balakrishnan Nair, 1969. Biochemical changes in relation to the breeding cycles of *Nausitora hedleyi* schepmman (Bivalvia: Teridinidae) Curr. Sci., 38(7) : 156-160.

Sasibhushana Rao, P., 1980. Pesticide toxicity to fish eggs and juveniles. Ph.D. Thesis, Annamalai University, India 210 pp.

Satyamurthi, S.T., 1956. Scaphopoda, Pelecypoda and Cephalopoda, Bull. Mad. Govt. Mus. The molluscs of krusodal Island. (In the Gulf of Mannar). New Sr., N.H. Sect. 1(2) pt.7, pp. 1-202, pls. 30.

Schimmel, S.C., J.M. Patrick, Jr. and J. Forester, 1977a. Toxicity and bioconcentration of BHC and lindane in selected estuarine animals. Arch. Environ. Contam. Toxicol., 6(2/3) : 355-363.

Scott, D. and C. Major, 1972. The effect of copper (II) on survival respiration and heart rate in the common blue mussel *Mytilus edulis* Biol. Bull. 143 : 679-688.

Sellmer, G., 1967. Functional morphology and ecological life history of the gem clam, *Gemma gemma* (Eulamellibranchiata: Veneridae) Malacologia, 5(2) : 137-223.

Shahul Hameed, P., 1984. Studies onthe crustalline style of some bivalve molluscs of Porto Novo coastal waters. Ph.D. Thesis, Annamalai University.

Shanmugam, A., 1987. Studies on *Pythia plicata* (Gray) (Gastropoda: Pulmonata: Elliobidae) from Pitchavaram mangroves, Ph.D. Thesis, Annamalai University, 154 pp.

Sivakumar, V., 1982. An environmental inventry of the tidal zone of the Vellar estuary. Ph.D. Thesis, Annamalai University, 152 pp.

Smith, D.A.S., 1971. Polymorphism and population density in *Donax rugosus* (Lumellibranchiata: Donacidae) J. Zool. London, 164 : 429-441.

Stakinken, D.M., 1978. Effects of uptake and discharge of petroleum hydrocarbon on the respiration of the soft-shell clam, *Mya arenaria.* J. Fish. Res. Bd. Can. 35 : 637-642.

Stekoll, M.S., L.E. Clement and D.G. Shaw 1980. Sublethal effects of chronic oil exposure on the intertidal clam. *Macoma balthica,* Mar. Biol., 57 : 51-60.

Stephen, A.C., 1928. Notes on the biology of *Tellina tennis* da costa. J. Mar. Biol. Ass. U.K., 15 : 327-341.

Stephen, A.C., 1929. Notes on the rate of growth of *Tellina tenuis* da costa in the Firth of Clyde, J.. Mar. Biol. Ass. U.K., 16 : 117-129.

Stickney, A.P., 1952. Studies on the gonads and the reproductive cycle of *Mya arenaria.*

Fourth annual conference on clam research. U.S. Fish and Wildlife Service, Boothbay Harbour, Marine 1-3, Ums.

Stirling, E.A., 1970. Some observations on the response of the benthic bivalve *Tellina tenuis* to pollutants. Technicoe Report. I.C.E.S.C.M. 1970/E:15.

Strickland, J.D.H. and T.R. Parsons, 1968. A practical hand book of seawater analysis. Ibid., 167: 311.

Strickland, J.D.H. and T.R. Parsons, 1972. A manual of sea water analysis (2nd edition) Bull. Fish. Res. Canada, No. 167: 310 pp.

Sreenivasan, P.V., 1980. Growth of the clam *Meretrix casta* (chemnitz) transplanted in the estuary. MPAI symp. coastal. Aquaculutre. Abstract No. 100.

Subramaniyan, B.R. T. Lingaraja and V.K. Venugopalan, 1979. Effect of low concentrations of DDT on the growth and production of marine diatom *Skeletonema costatum* (Grew), Curr. Sci., 48 : 226-228.

Sundaramurthy, S., 1987. Taxonomical developmental and larval bioassay studies in the mangrove crab Neoepisesarma (Murradium) *Tetragonum Fabricius* 1798. (Grapsidae, Brchyure, Decapoda).

Sunilkumar Mohamed, K., 1981. A review on lethal and sublethal effects of petroleum contaminants on marine shell fishes. M.Sc., Thesis, Annamalai University.

Tanabe, S. and R. Tatsukawa, 1980. Chlorinated hydrocarbons in the north pacific and Indian oceans. J. Oceangr. Soc. Japan, 36 : 217-226.

Tarrant, K.R., and J.O.G. Tatton, 1968. Organochlorine pesticides in rain water in the Bristish Isles. Nature Lond., 219 : 725-727.

Tarzwell, C.M., 1963. In: Pesticides their use and effect (New York state Joint Legislative Committe on Natural Resources, Albany, New York, U.S.A., 30).

Thivakaran, G.A., 1988. Studies on the Littrinids *Littorina quadricentus* (Philippi) and *Nodilittorina pyramidalis* (Quoy and Gaimard, 1833) (Gastropods: Prosobranchia: Littorinidae) from the Tranquebar rocky shore (South east cost of India) Ph.D. Thesis, Annamalai University.

Thorson, G., 1957. Communities (sublittoral or shallowshelf) pp. 461-534. In Hedgpeth, J.W. (Ed). Treatise on Marine Ecology and Paleocology. Mem. Geol. Soc. Am., No. 67 (1), 1296 pp.

Thurberg, F.P., 1980. The use of physioloigcal techniques in monitoring pollution. A consideration of its problems and current research. Rapp. P.V. Reun. Cons. Perm. Int. Explor. Mer. 179, 82-87.

Thurberg, F.P., Cable, W.D., Dawson, M.A., MacInnes, J.R. and D.R. Wenzloff, 1975. Respiratory response of larval juvenile and adult surf clams, *Spisula solidissima* to silver respiration of marine organisms.TRIOGOM Publ. S. Portland. Me.

Thurberg, F.P., A. Calabrese and M.A. Dawson, 1974. Effects of silver on oxygen consumption of bivalves at various salinities. In: Pollution and physiology of marine organisms (F.J. Vernberg and W.B. Vernberg eds :,) Academic Press, New York, pp. 67-78.

Thurberg, F.P., M.A. Dawson and R.S. Collier, 1973. Effects of copper and cadmium on osmoregulation and oxygen consumption in two species of estuarine crabs. Mar. Biol., 23 : 171-175.

Topping, F.L. and J.L. Fuller, 1942. The accomodation of some marine invertebrates to reduced osmotic pressures. Biol. Bull., 82 : 372-389.

Tucker Abbott, 1954. American sea shells. Van Nostrand Company. INC Princeton New Jersey. 1 : 24-29.

Underwood, A.J., 1979. The ecology of intertidal *gastropods. Advances in Marine Biology, 16*; 111-210.

Venugopalan, V.K., 1955. Studies in the anatomy and physiology of *Sanguinearia diphos* (Gmelin) M.Sc., Thesis, Annamalai University.

Verma, S.R., S.K. Bansal and R.C. Dalela, 1977. Bio-assay trials with *a* few organic biocides on freshwater fish, *Labeo rohita,* Indian J. Environ. Hlth., 19 : 107-115.

Verma, S.R., S.P. Gupta and M.P. Tyagi, 1975. Studies on the toxicity of lindane on *Colisa fasciatus* (Part I : TLM measurements and histological changes in certain tissues) Morphol. Jahrb. 121 (1) : 38-54.

Vijayalakshmi, G.S., 1973. Hydrobiological studies in the Vellar estuary. Ph.D. Thesis, Annamalai University, 183 pp.

Vijayalakshmi, G.S. and V.K. Venugopalan, 1973. Diurnal variation in the physico-chemical and biological properties in Vellar estuary. Indian J. Mar. Sci., 1 : 125-127.

*Vijayamadhwan, K.T. and T. Iwai, 1975. Bull. Jap. Soc. Fish., 41 : 634-639.

Waugh, G.D. and Ansell, A. 1956. The effect on oyster spatfull of controlling barnacle settlement with DDT, Ann. Appl. Biol. 44 : 619-625.

Wildish, D.J., 1977. Factors controlling marine and estuarine sublittoral macrofauna. Helgo lander. Wiss. Meeresunters. 30 : 445-454.

Wilcott, T.G., 1973. Physiological ecology and intertidal zonation in limpets (ACMAEA) : A critical look at limiting factors. Biol. Bull., 145 : 389-422.

Wolff, W.S., 1973. The estuary as a habitat. An analysis of data on the soft bottom macrofauna of the estuarine area of the River Rhine, Meuse and Scheldt, Zoologische. Verhandelind gen, No. 126 : 3-242.

Yonge, C.M., 1923. Studies on the comparative physiology of digestion. The mechanism of fedding, digestion and assimilation in the lamel1ibranch Mya. Brit. J. Ex. Biol., 1 : 15-63.

Yonge, C.M., 1926a. Digestive diverticula in lamelibranchs. Trans. R. Soc. Edinburgh, 54 : 703-718.

Yonge, C.M., 1926b. Structure and physiology of the organs of feeding and digestion in *Ostrea edulis* J. Mar. Biol. Assoc. U.K., 14 : 295-386.

Yonge, 1949. On the structure and adaptations of the Tellinacea deposit feeding eulamellibranchia, Phil. Trans. Roy. Soc., Biol. Ser. 234 : 29-76.

Yonge, D.K. and D.C., Rhoads, 1971. Animal sediment relatins in Cape Cod Bay. Massachusetts. I. A transect study. Mar. Biol., 11(3), 242-254.

Woodwell, G.M., C.M. Wurster and P.A. Isaacson. 1967. DDT residues in an East coast estuary. A case of biological concentration of a persistent insecticides. Science, 156 : 821-824.

*Zaroogiam, G.E., 1980. Mar. Biol., 58 : 275

ADDENDUM

Brock, V., 1980. Notes on relations between density settling and growth of two sympatric cockles, *Cardium edulle* (L) and *C. glaucum* (Bruguiere). Ophelie. Suppl., 1 : 241-248.

Coe, W.R., 1947. Nutrition, growth and sexuality in the pismo clam *(Tivela stultorum)* J. Exptl. Zool. 104 : 1-24.

Coe, W.R., 1955. Ecology of the boan clam *Donax gouldrin* on the east coast of Southern California. Ecology, 36 : 512-514.

Goshima, S., 1982. Population dynamics of the soft clam *Mya arenaria* L., with special reference to its life history pattern. Publ. Amatensa Mar. Biol. Lab., 6 : 119-165.

Kasinathan, S. and S.R. Vijayaraghavan, 1964. Effect of Insulin on the Aminonitrogen content of free amino acids in the blood and foot of *Meretrix casta* (Chemnitz). Curr. Sci. 33(9) : p 276-277.

Morton, B., 1974. Some aspects of the biology, population dynamics and functional morphology of *Musculis ta senhansia.* Benson (Bivalvia: Mytilidae) Pacific Science. 28 : 19-33.

Newcombe, C.L., 1936. A comparative study of the abundance and the rate of growth of *Mya arenaria* L. in the Gulf of St. Laurence and Bay of Fundy regions. Ecology, 17 : 418-428.

Orton, J.H., 1926. On the rate of growth of *Cardium edule.* Part I. Experimental observations. J. mr. biol. Ass. U.K., 14 : 239-279.

Penzias, L.P., 1969. *Tellina martinicensis* (Mullusca: Bivalvia) Biology and Productivity. Bull. Mar. Sci., 19(3) : 568-579.

Ramamoorthi, K., 1954. A preliminary study on the hydrology and fauna of the Vellar estuary (South Arcot). Proc. Indo-Pacific Fish. Counc. Symp., 9, Bangkok.

Rockford, D.J., 1951. Studies in Australian estuarine hydrology. I. Introductory and comparative features. Adust. J. Mar. Freshw. Res., 2 : 1-166.

Salzwedel, H., 1980. Energy budget for two populations of the bivalves. *Tellina fabula* in German Bight. Veroff. Inst. Meeresfursch. Bremerh. 18 : 257-287.

Segerstrale, 1960. Investigation on Baltic populations of the bivalve *Macoma balthica* (L). Part 1. Studies on recruitment and its relation to depth in Finfish coastal waters. 1922-1959. Age and growth. Commentat. Biol., 23 : 1-72.

Srihari, C.P., 1975. Studies on *Tellina rubra* (Mollusca: Bivalvia) occuring in the Vellar estuary. M.Sc., Thesis, CAS in Marine Biology, Annamalai University.

* Not referred in the original.

INDEX

A

B

C

D

E

F

R

S

T

✦✦✦

Pl. 1. Collection area–(Vellar estuary).

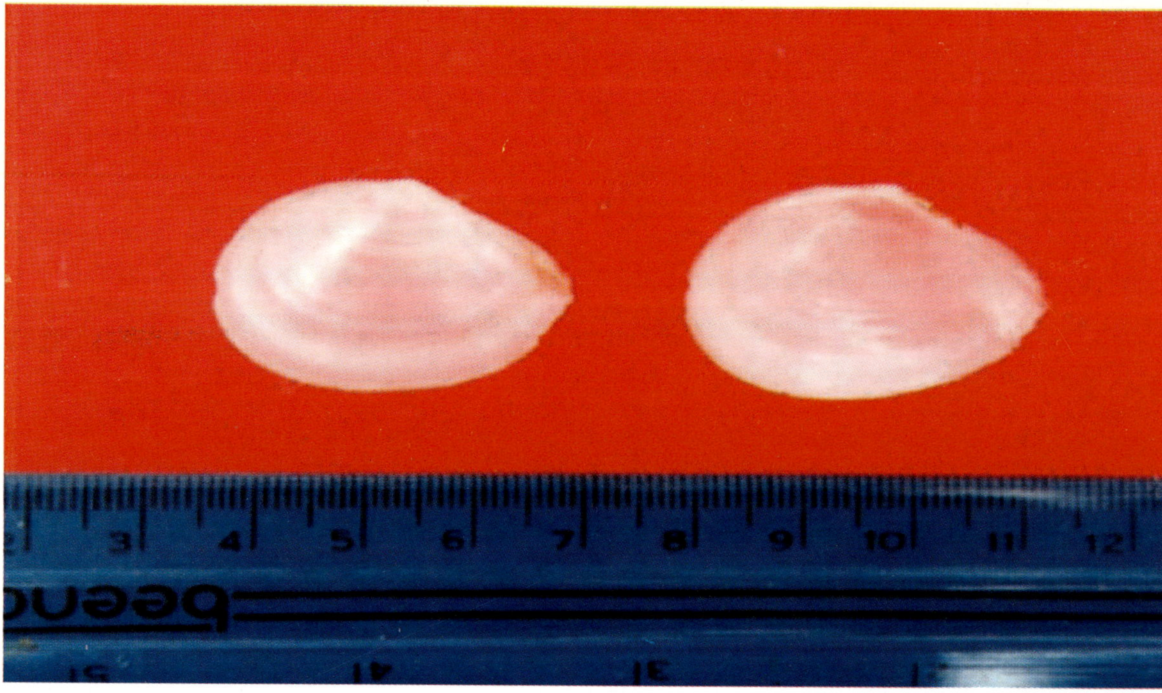

Pl. 2. *Tellina ala.*

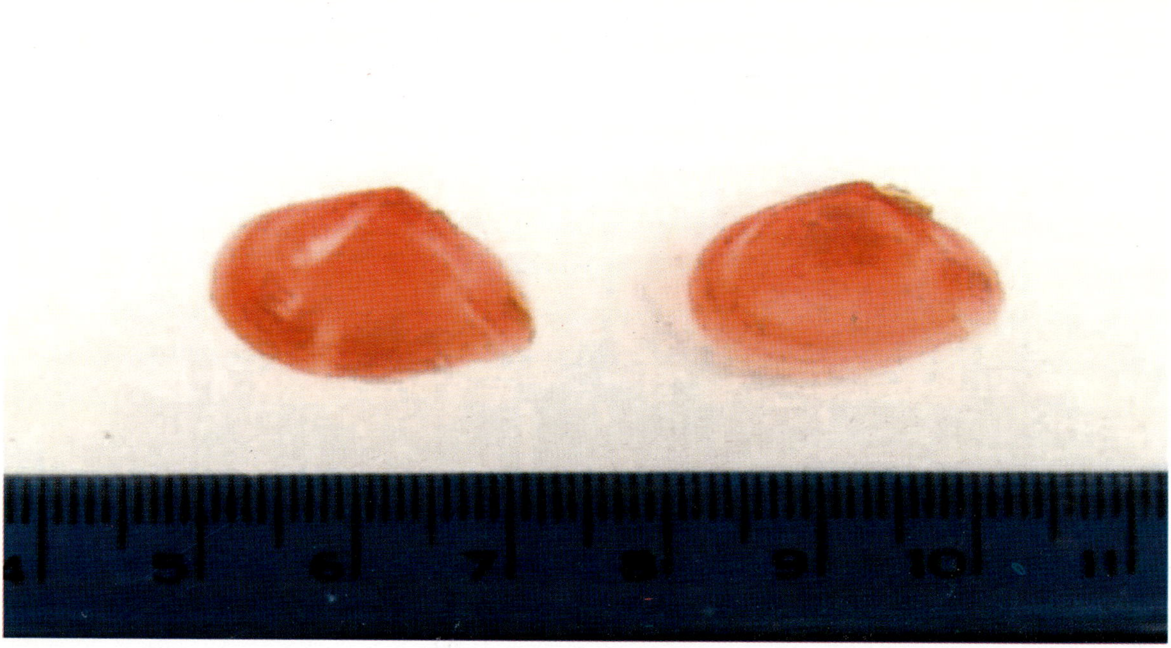

Pl. 3. *Tellina cuspis.*

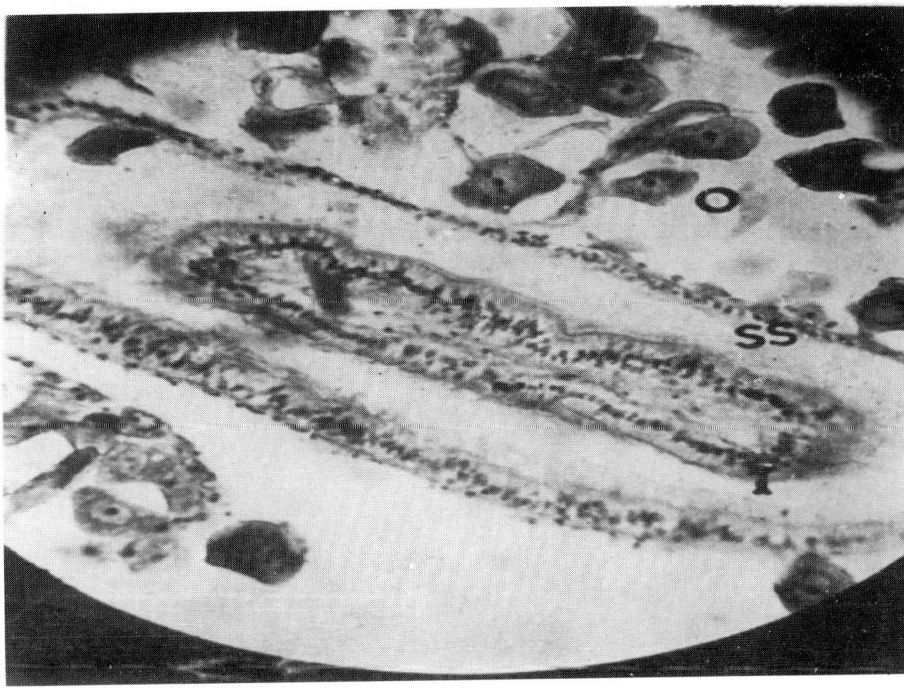

Pl. 4. T.S. through style sac and Intestine. X 320.

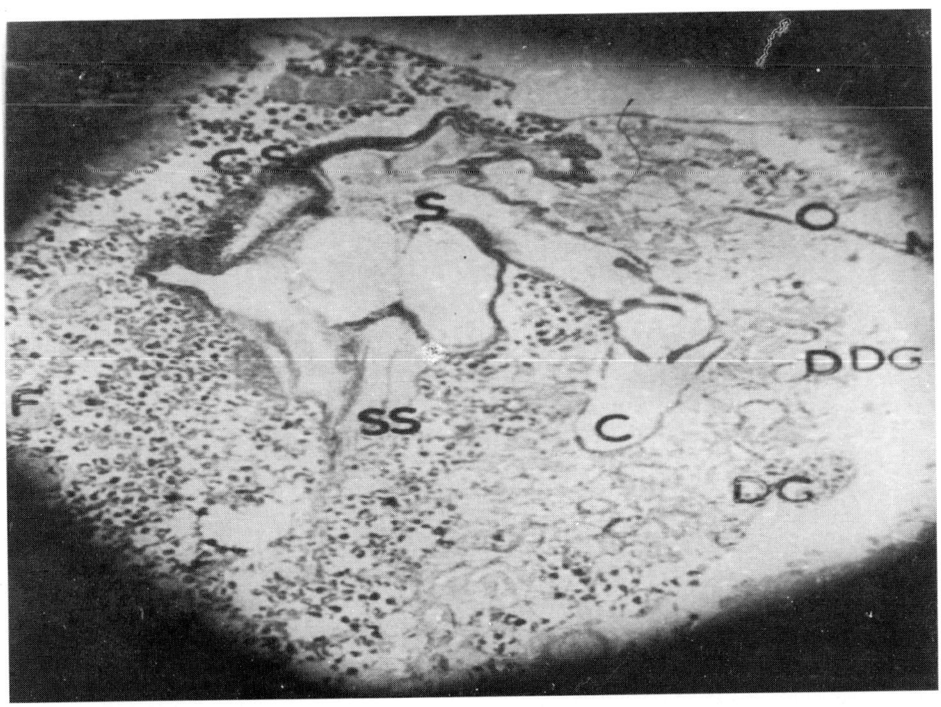

Pl. 5. T.S. through the stomach region. X 30.

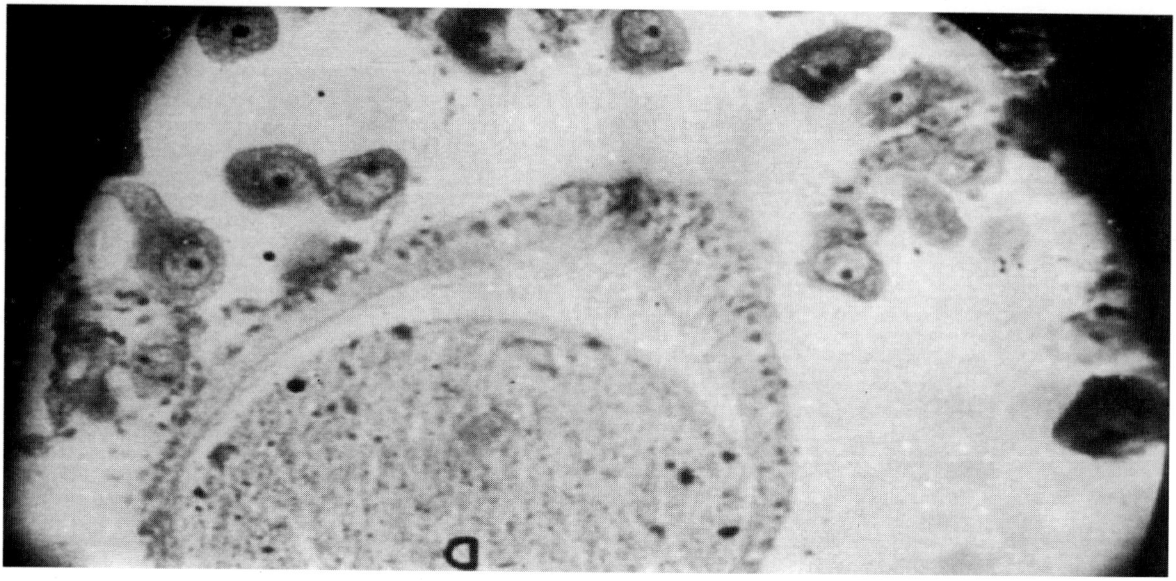

Pl. 6. Section passing through the intestine showing faecal matter. X 320.

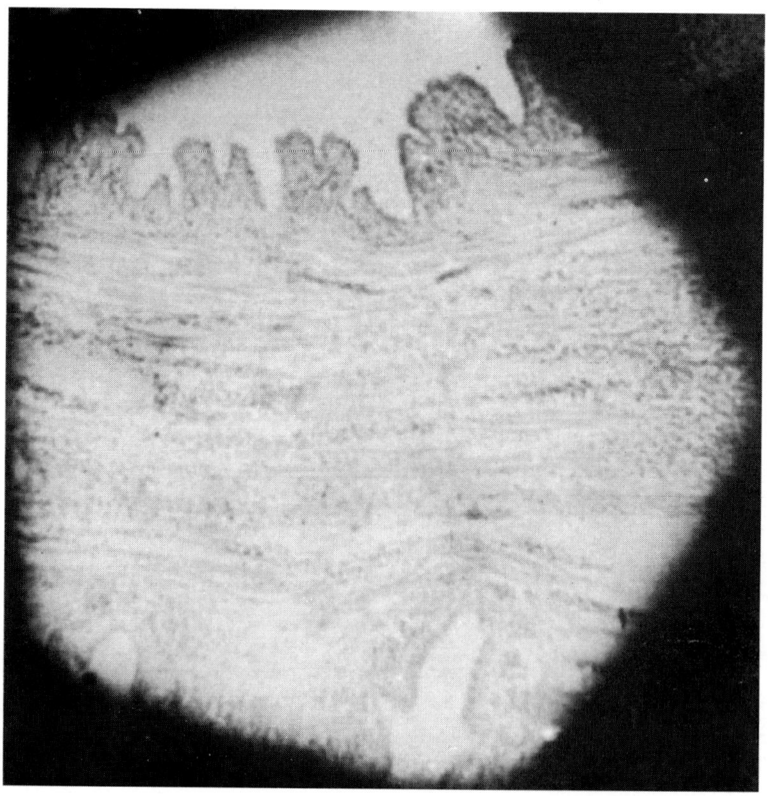

Pl. 7. T.S. through ctenidium. X 30.

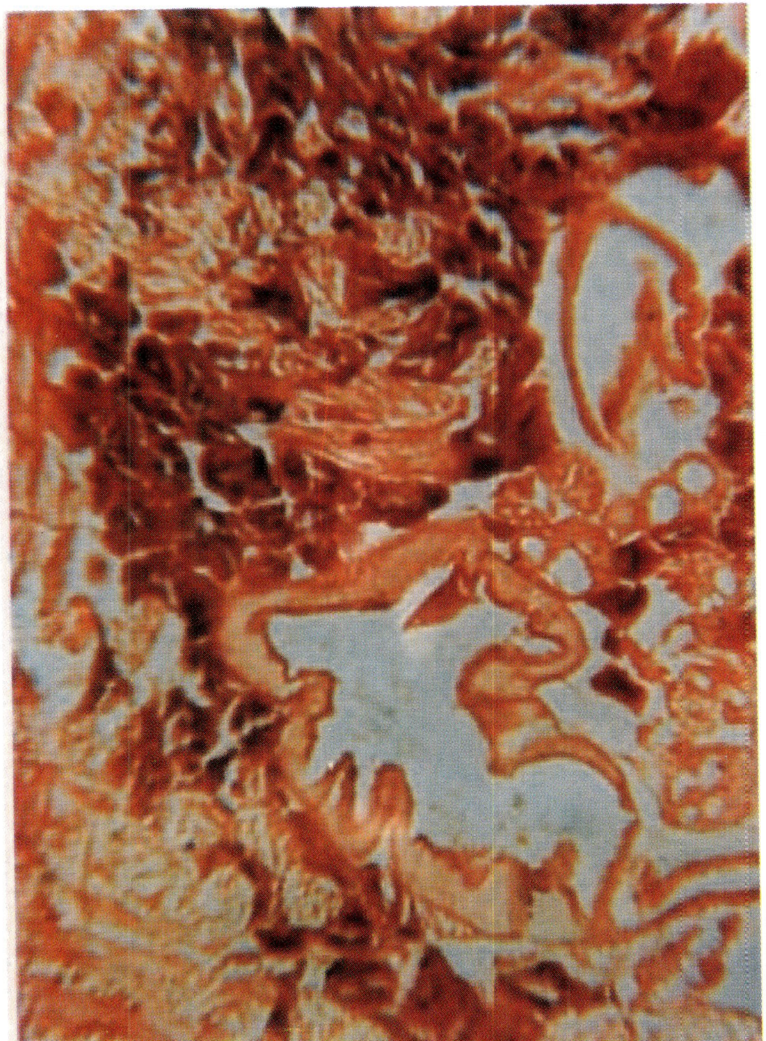

Pl. 8. C.S. of rectum. X 100

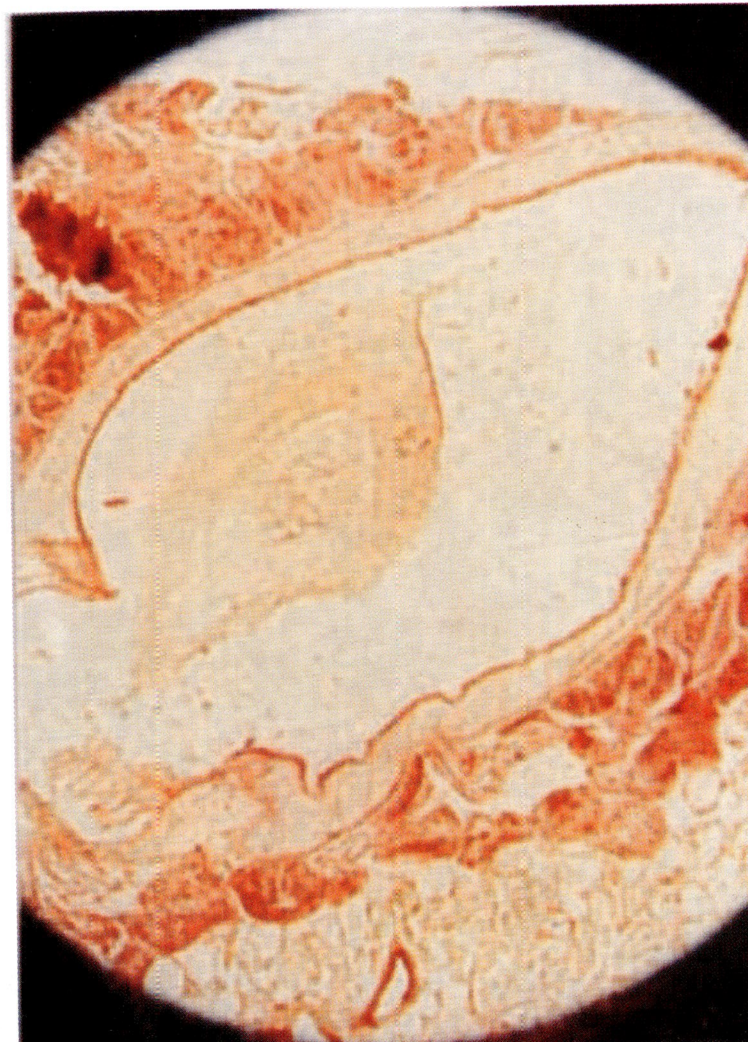

Pl. 9. C.S. of stomach. X 400

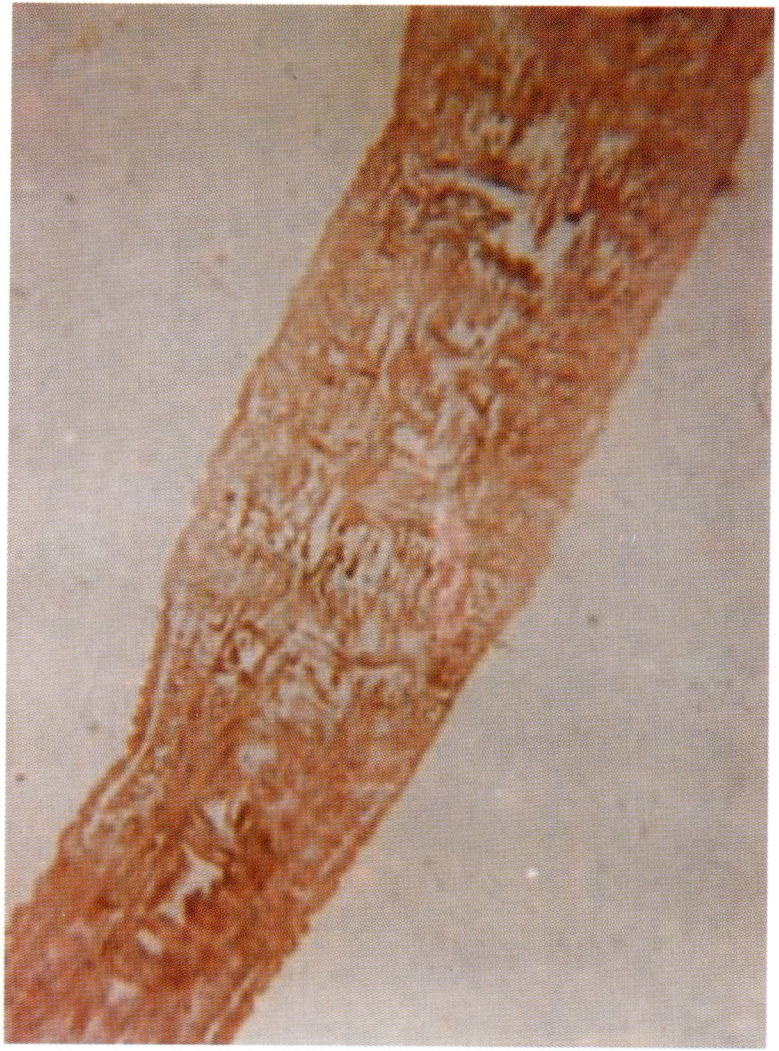

Pl. 10. T.S. of the foot. X 100

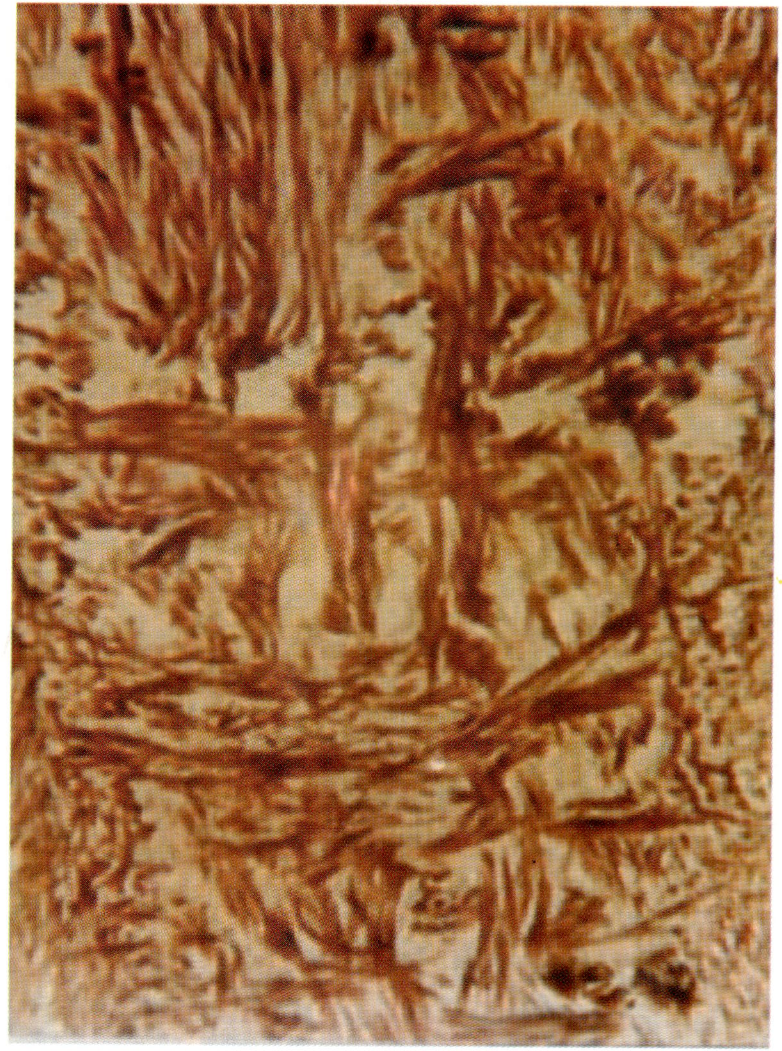

Pl. 11. T.S. of the foot. X 400

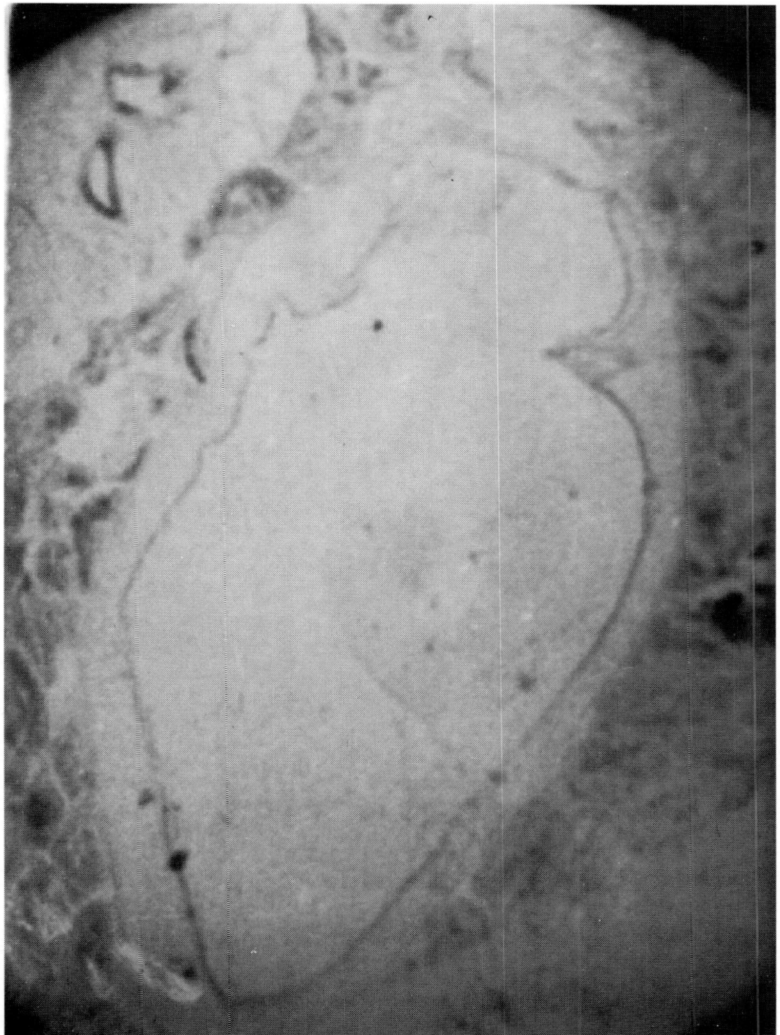

Pl. 12 C.S. of stomach. X 50

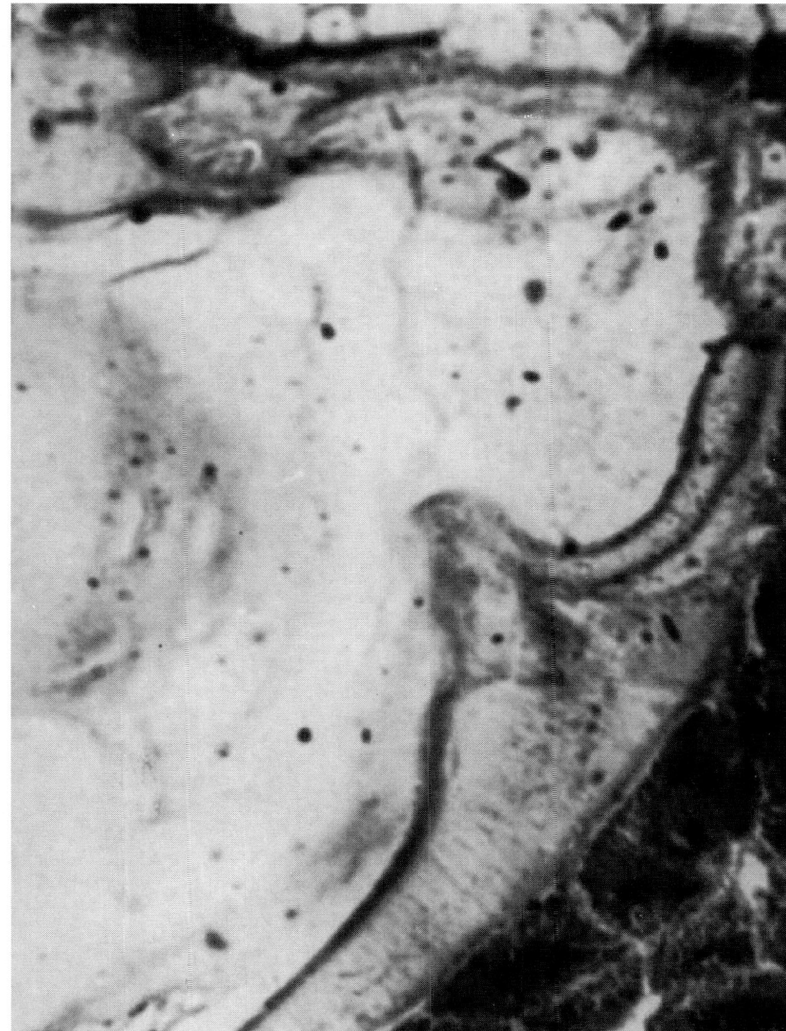

Pl. 13 Portion of stomach wall elongated. X 100

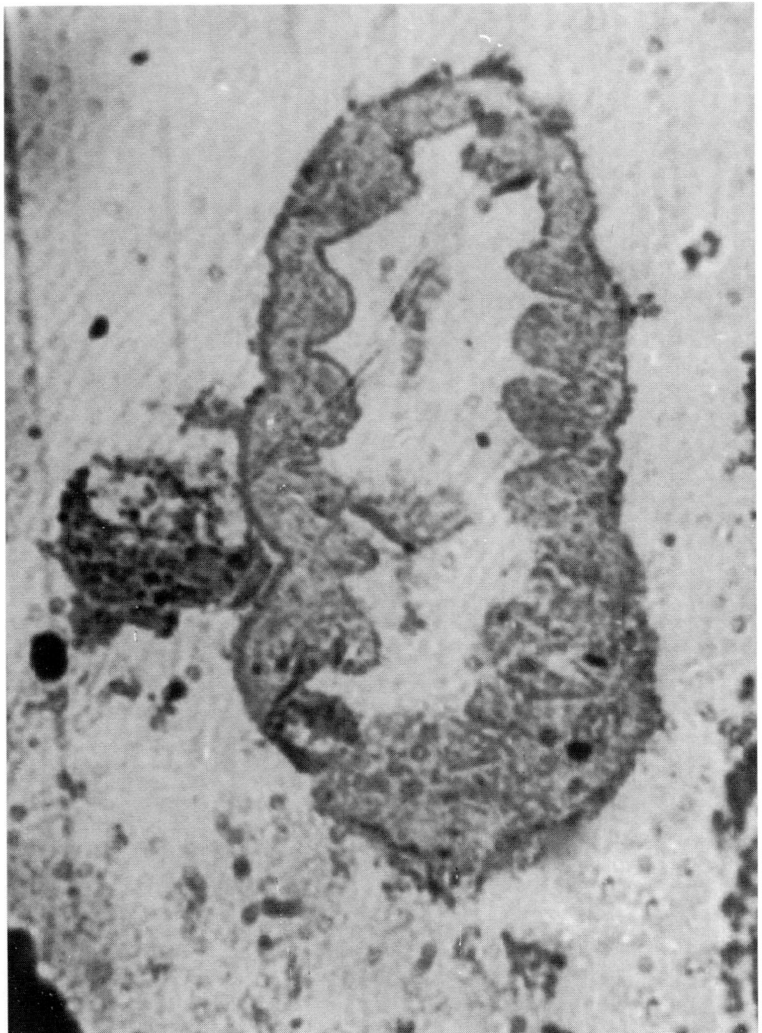

Pl. 14. C.S. of rectum. X 30

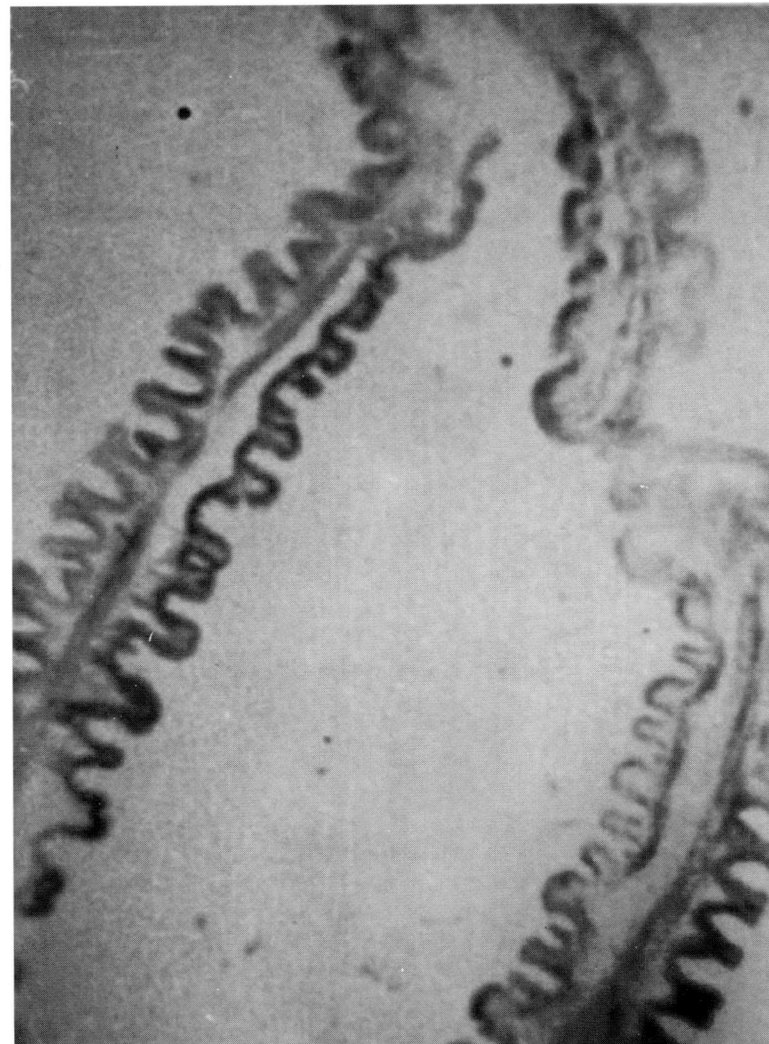

Pl. 15. L.S. of Siphon. X 100

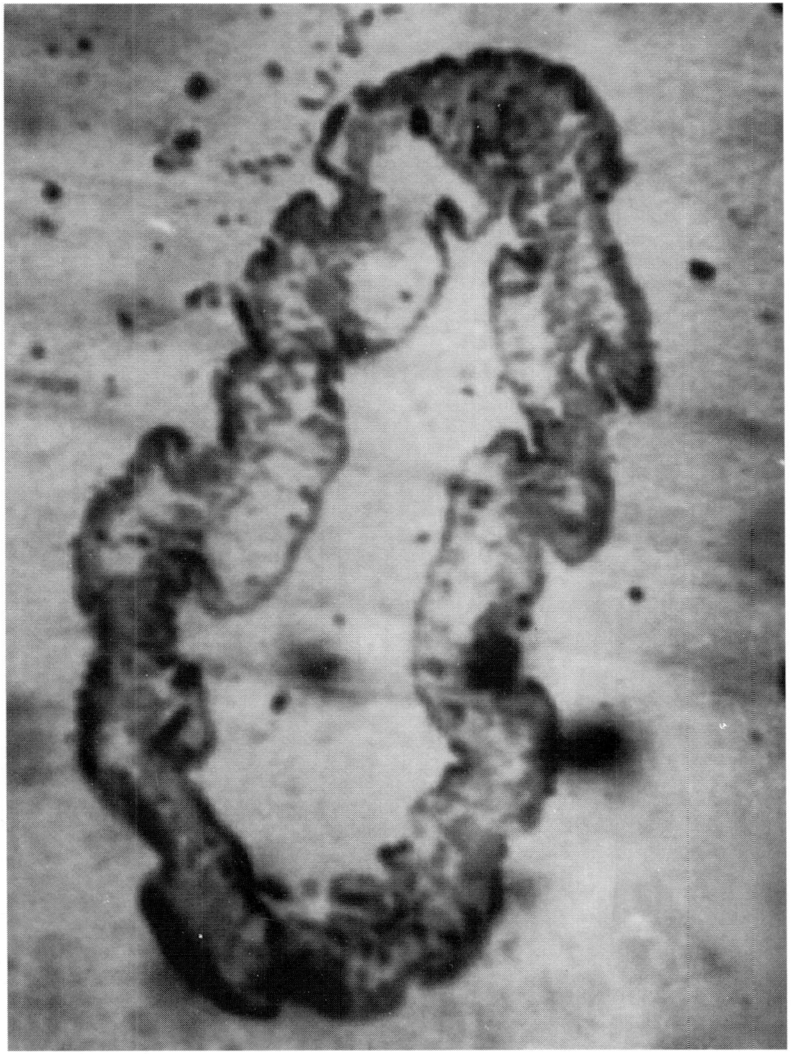

Pl. 16. C.S. of Siphon. X 30

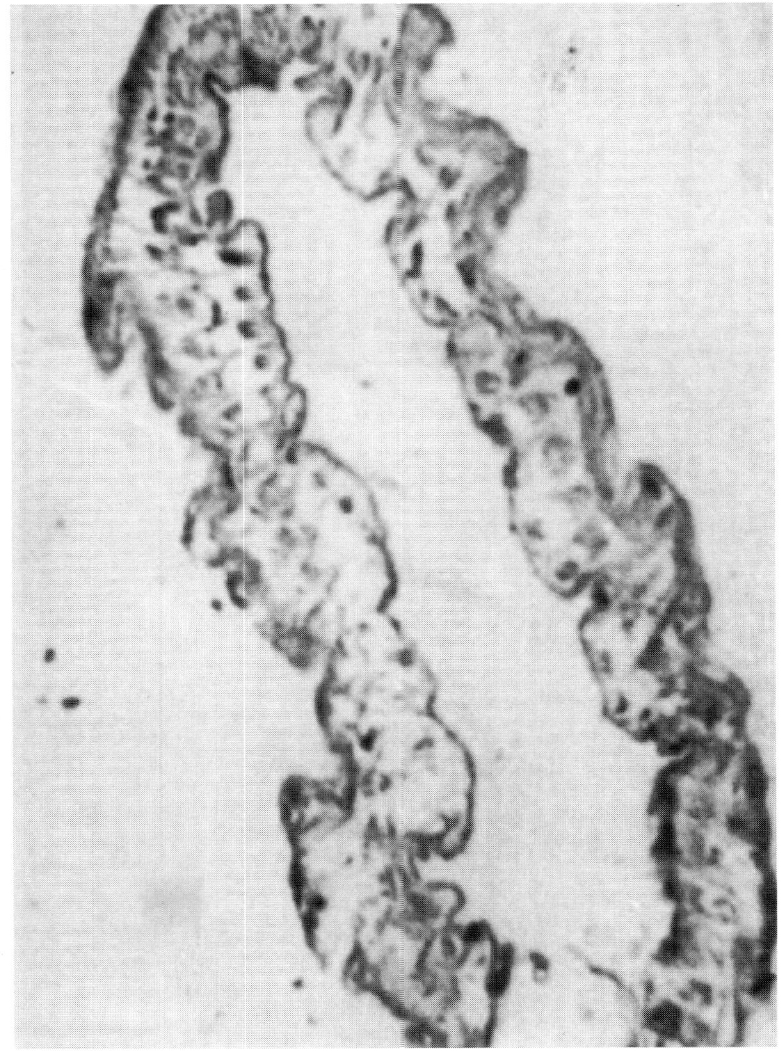

Pl. 17. C.S. of Siphon. X 100

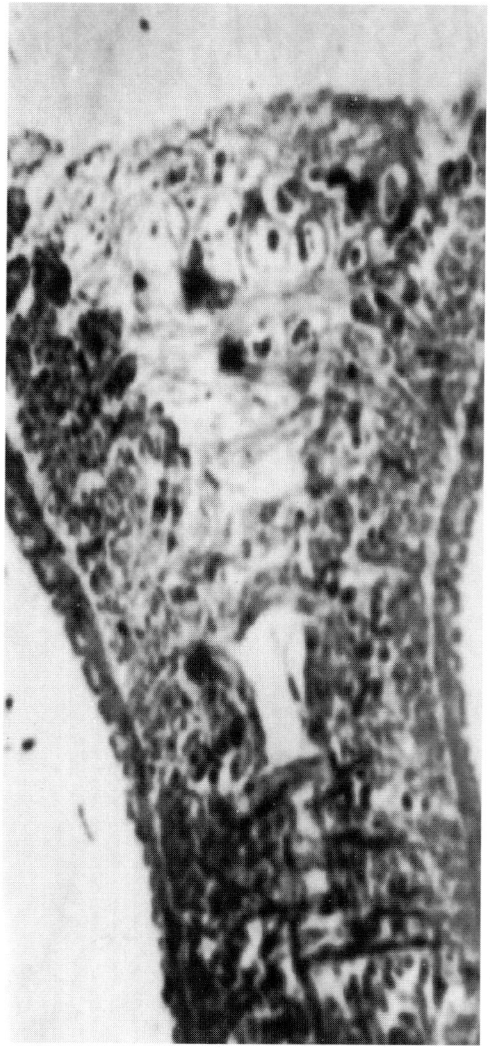

Pl. 18. T.C. of foot. X 100

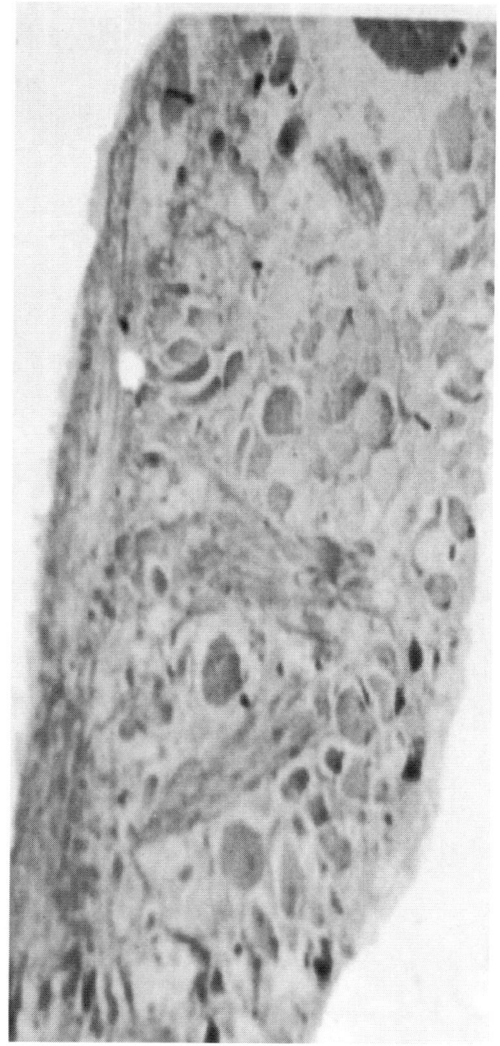

Pl. 19. C.S. of Occyte. X 50

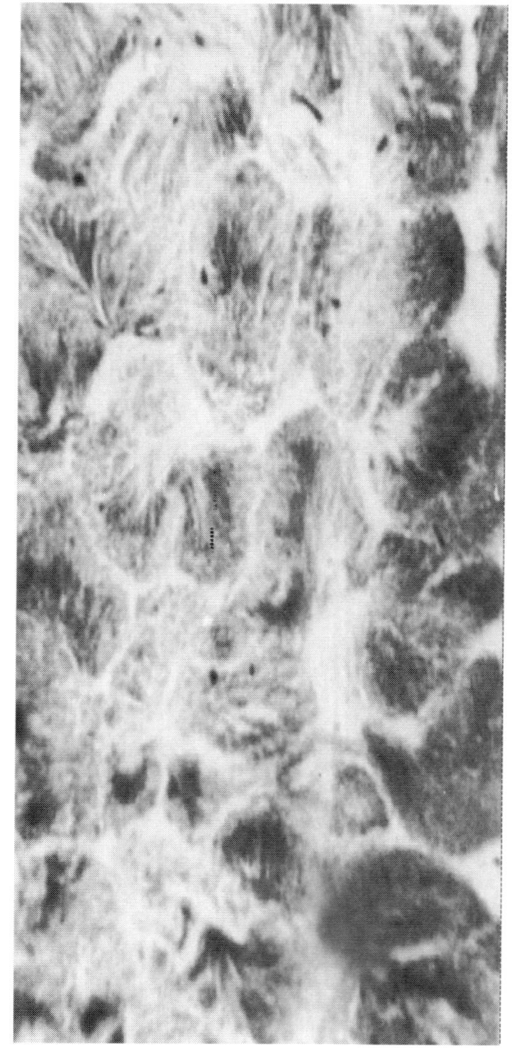

Pl. 20. C.S. of Testis. X 50

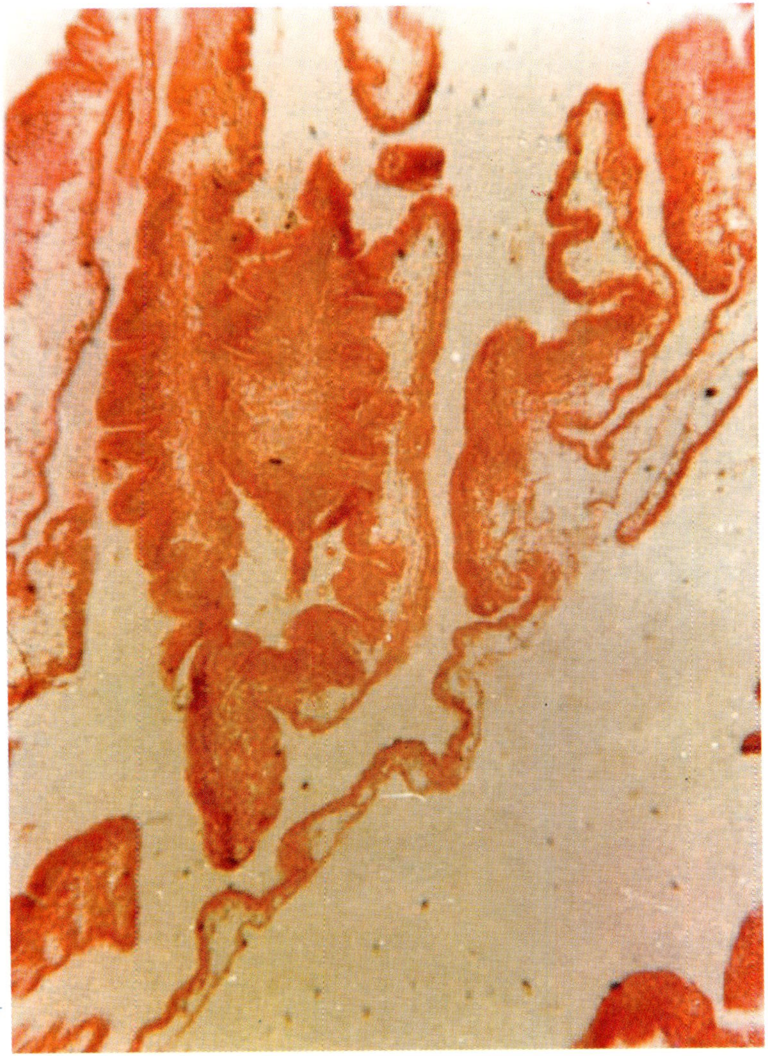

Pl. 22. C.S. through gill. X 100

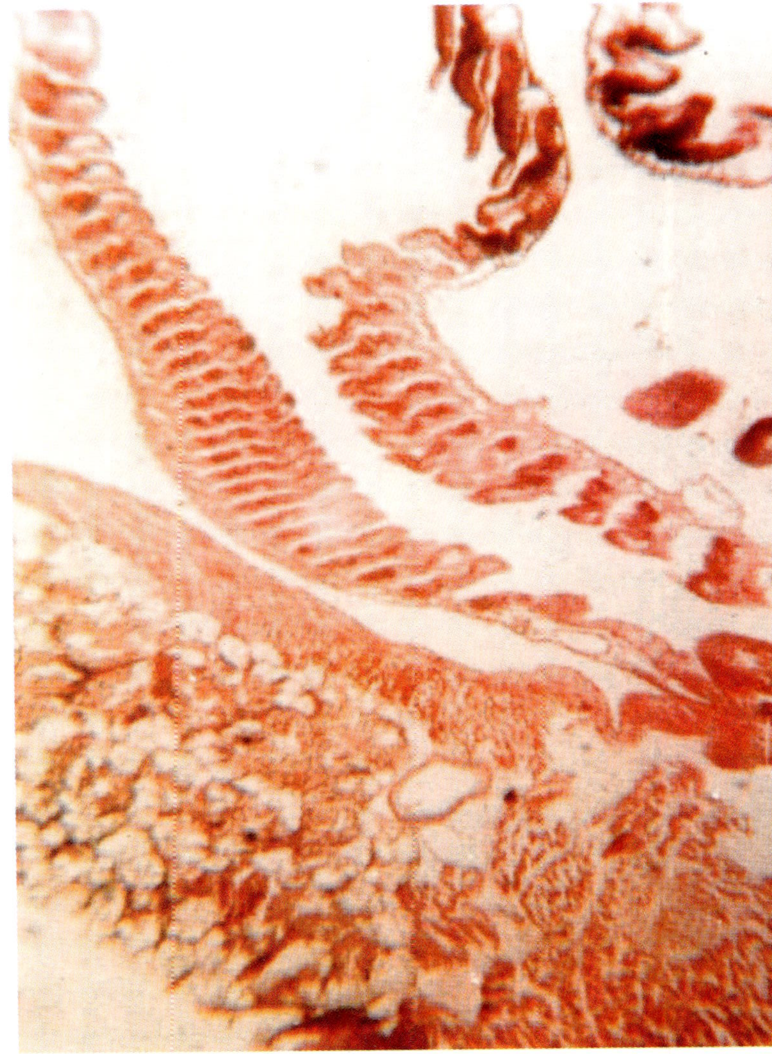

Pl. 21. T.S. through gill and digestive diverticula. X 100

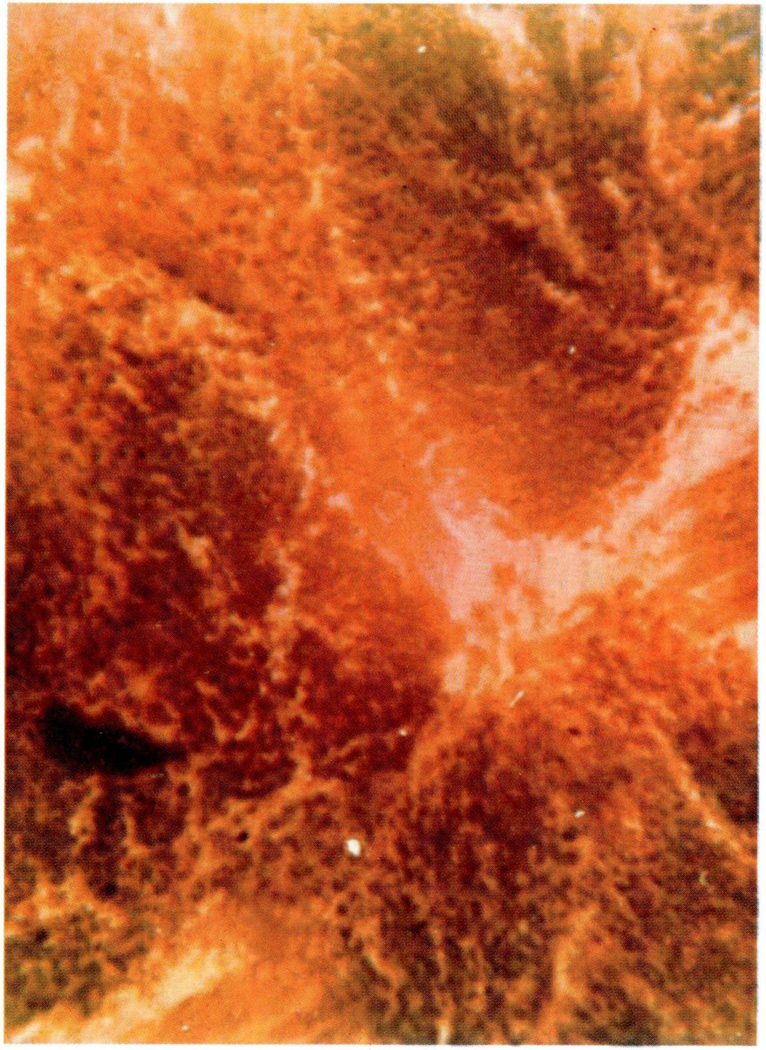

Pl. 24. C.S. through Testis. X 400

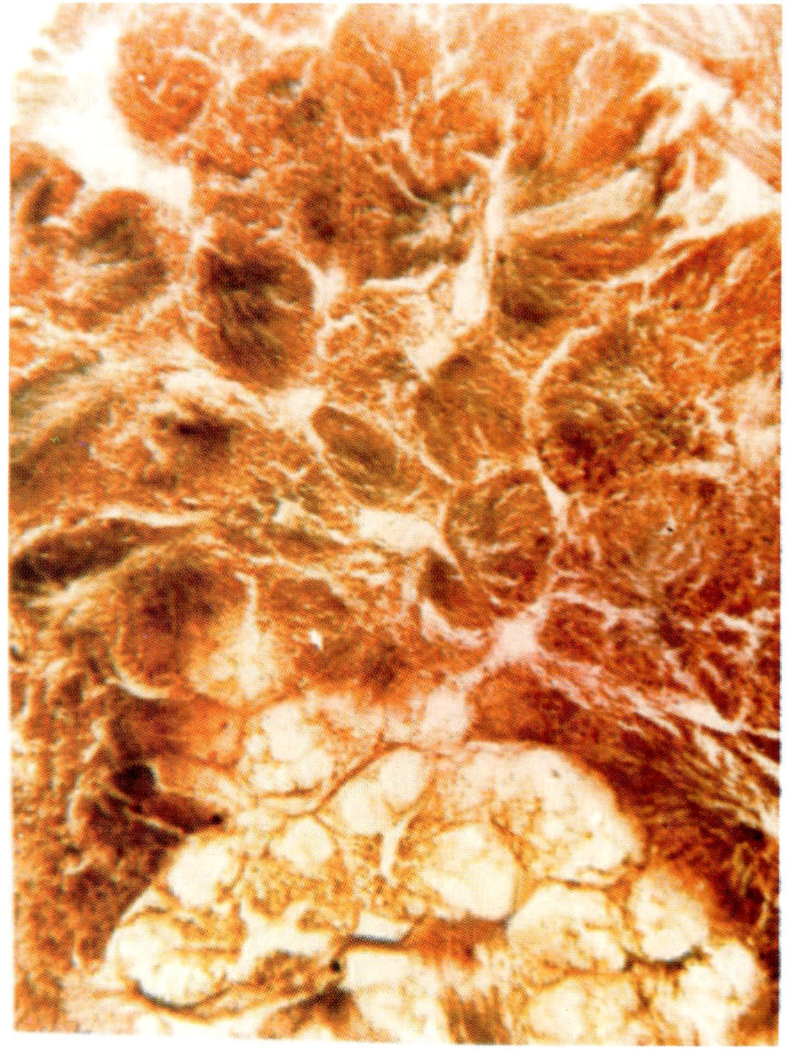

Pl. 23. C.S. through Testis. X 100

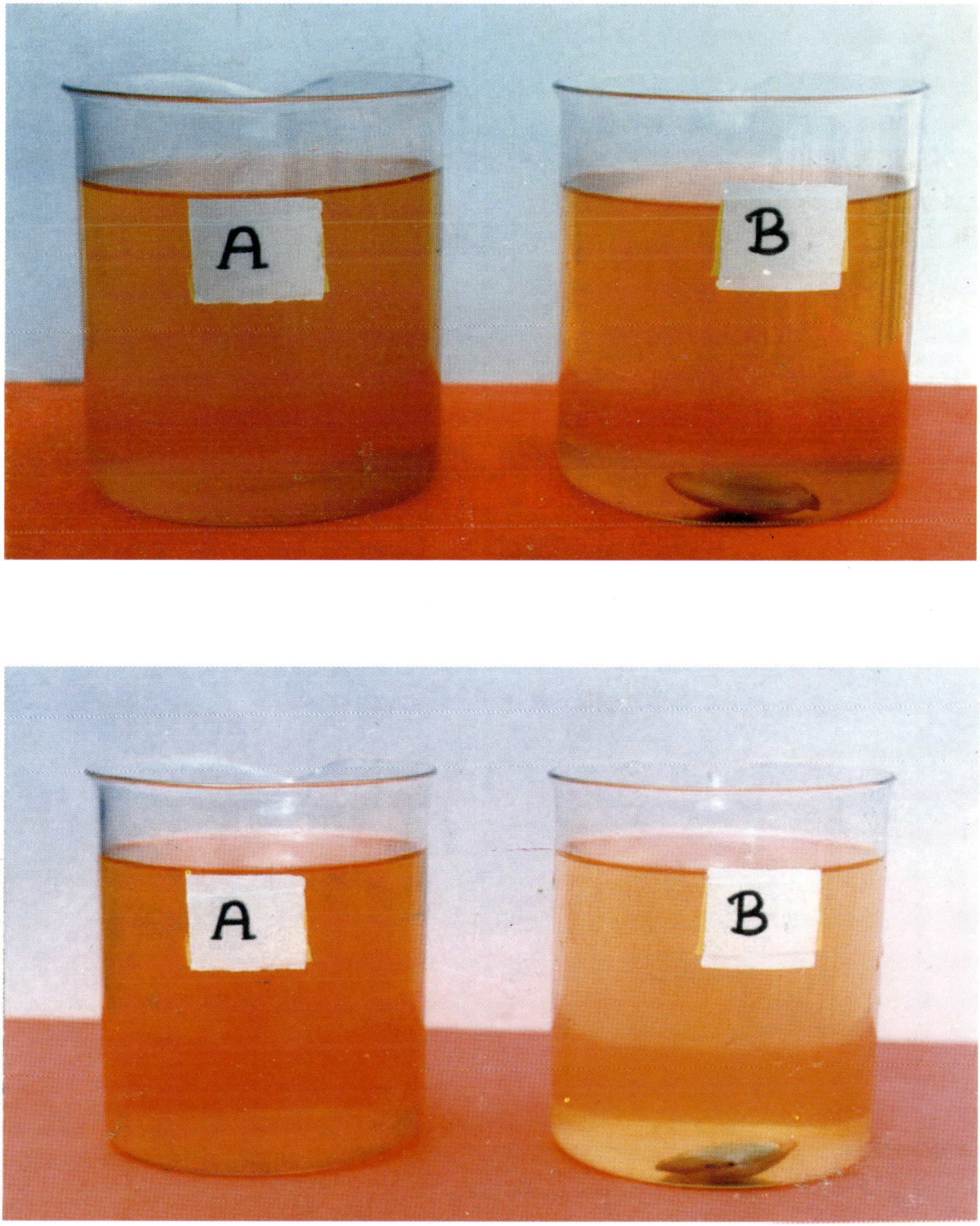

Pl. 25 A, B. *T. ala* and *T. cuspis* Rate of water filtration at initial stage (0 hour)
Rate of water filtration at final stage (after 1 hour)

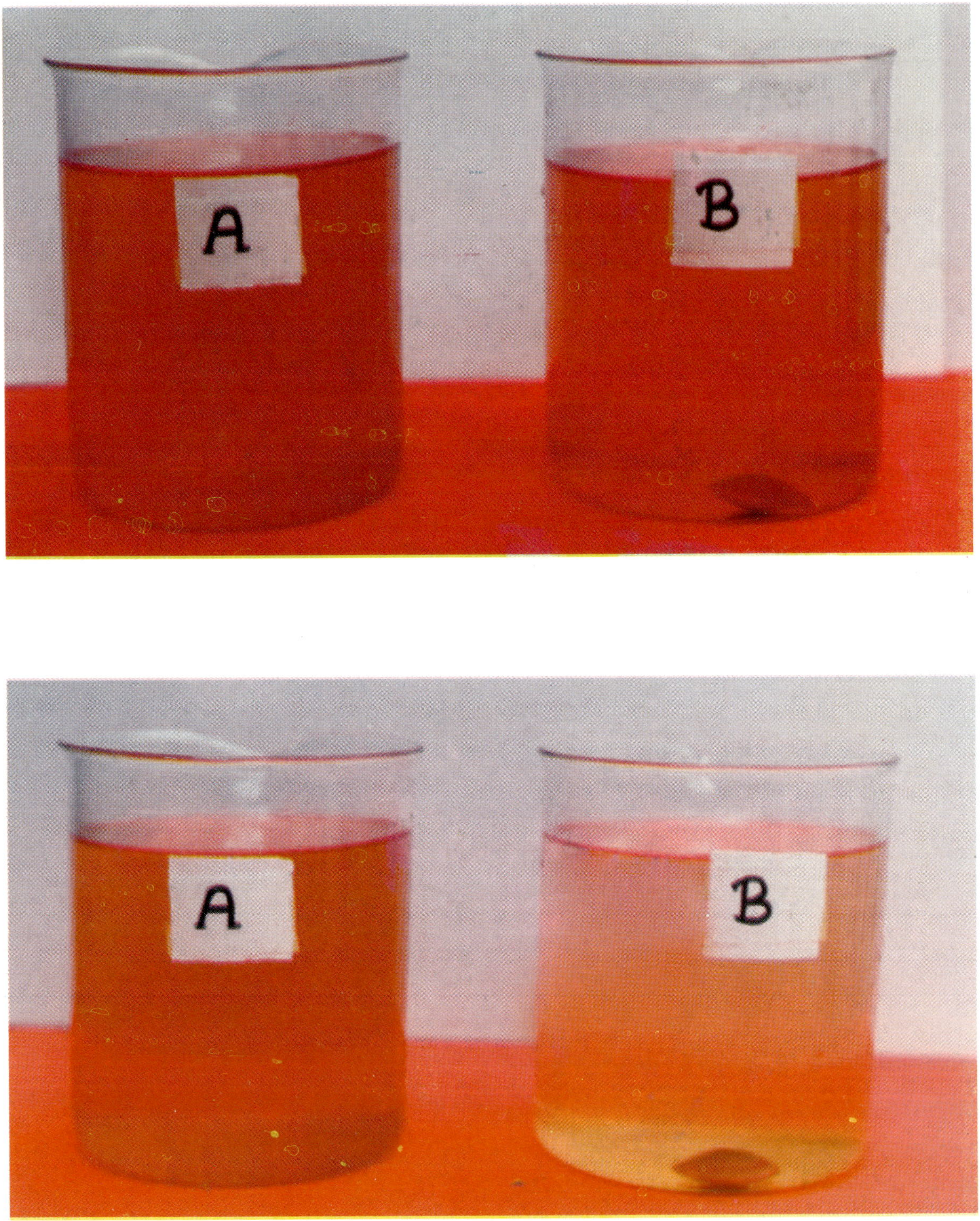

Pl. 25 C, D. *T. ala* and *T. cuspis* Rate of water filtration at initial stage (0 hour)
Rate of water filtration at final stage (after 1 hour)

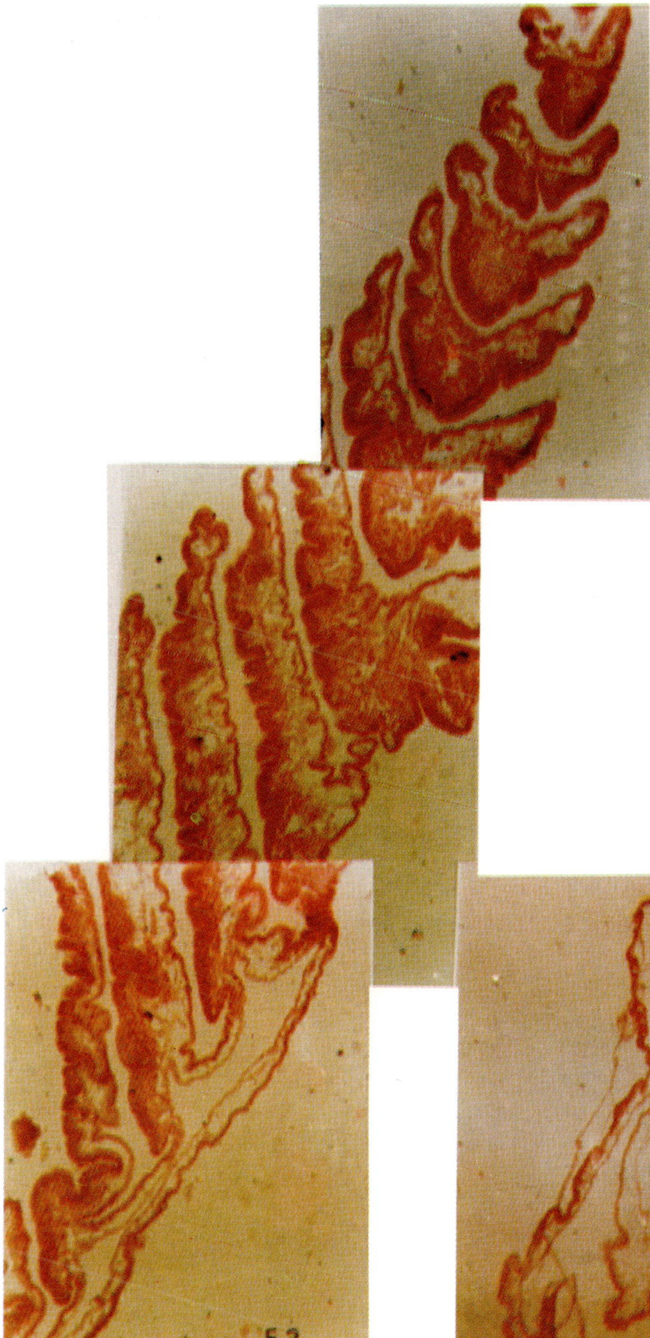

Pl. 26 A. Control gill of *T. ala*. H and E. X 100

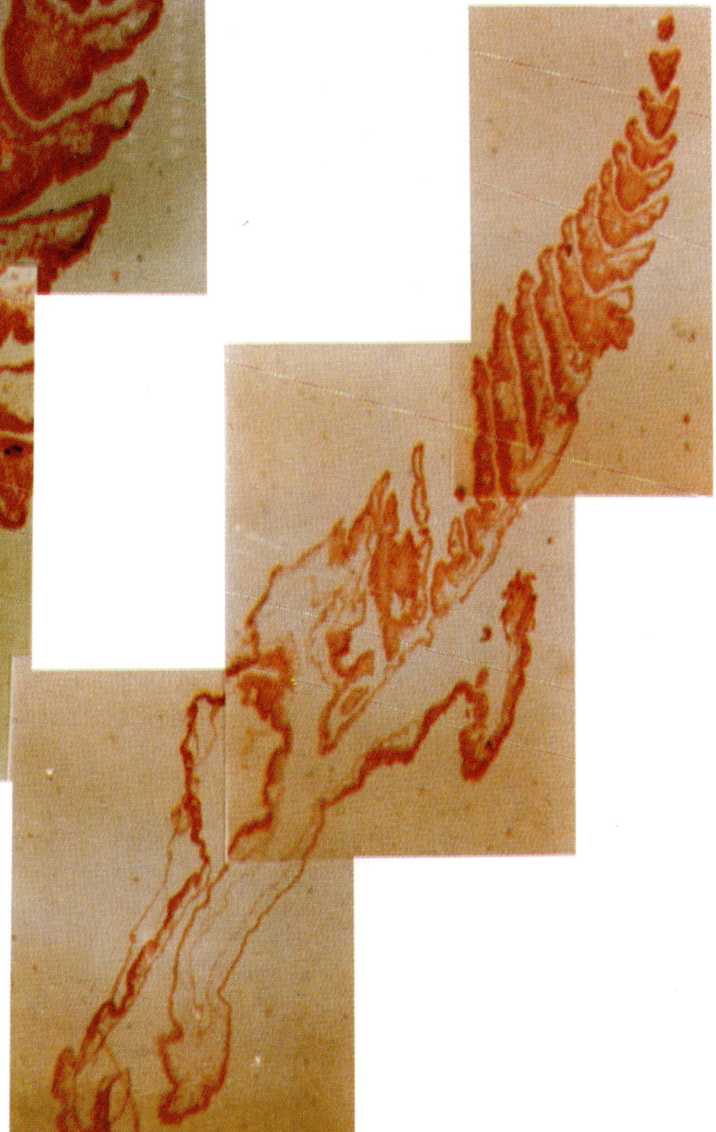

Pl. 26 B. Control gill of *T. ala*. H and E. X 400

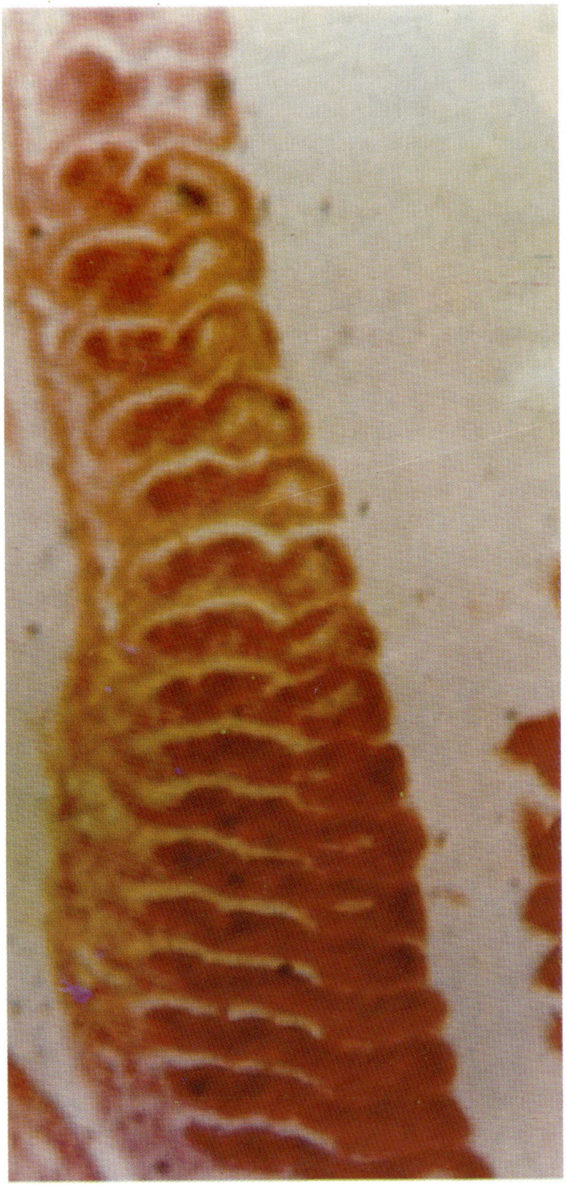

Pl. 27 A. A portion showing control gill of *T. ala.* X 100

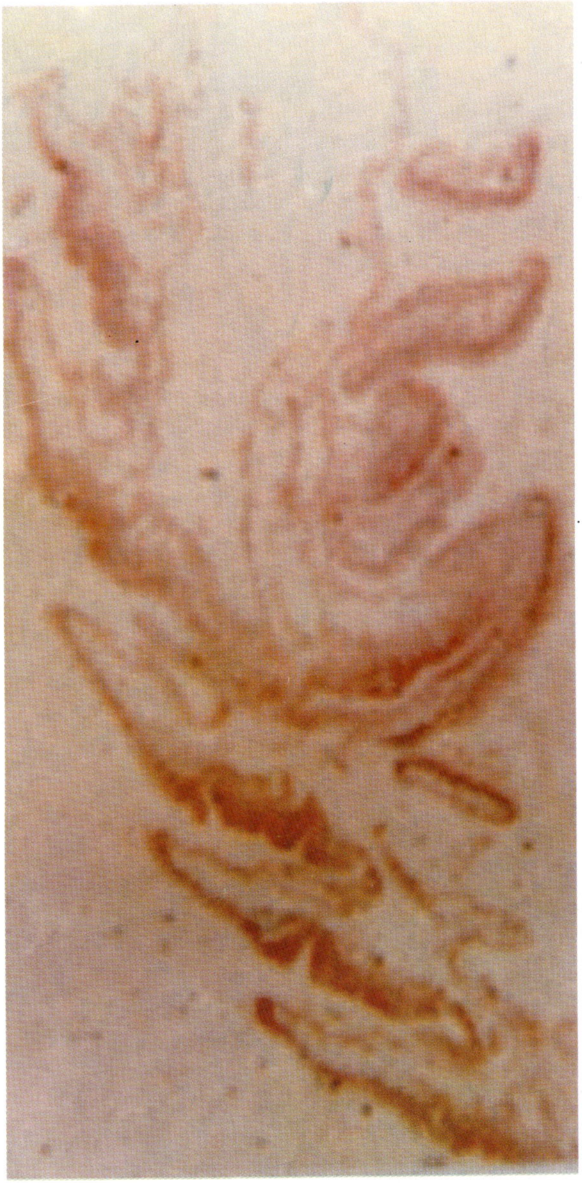

Pl. 27 B. Gill of *T. ala* exposed to acute concentration (10 ppb) of endosulfan. H and E. X 500.

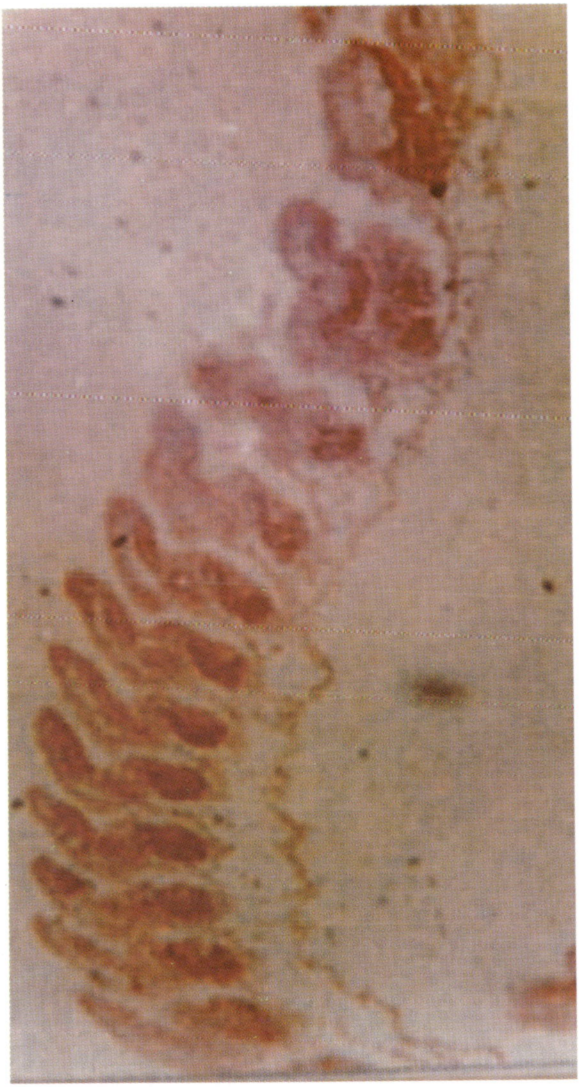

Pl. 27 C. Gill of *T. ala* exposed to sublethal concentrations (2 ppb) of endusolfan. H and E. X 100.

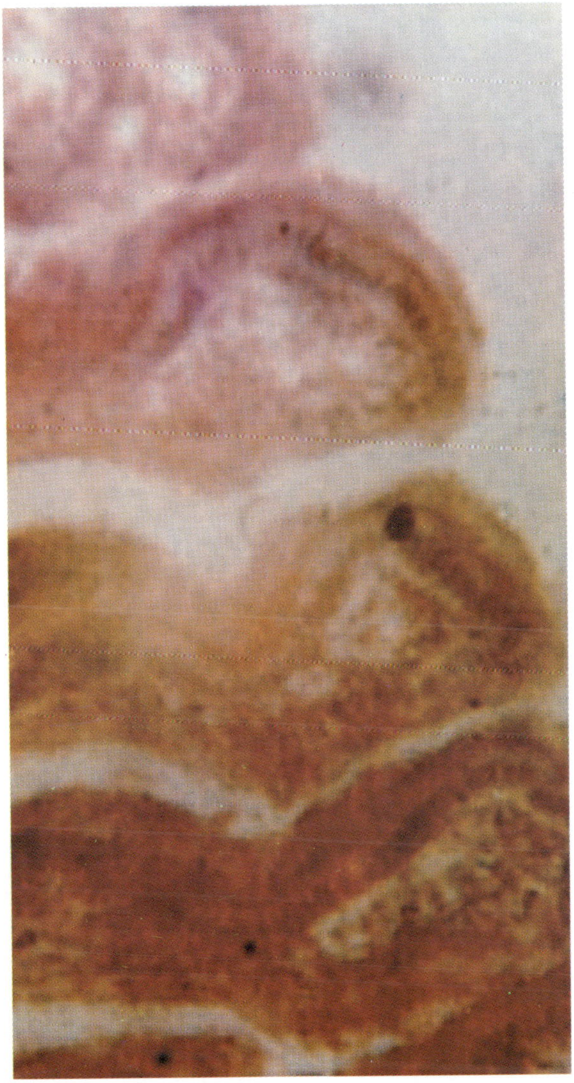

Pl. 27 D. Control Gill a portion enlarged view. X 400.

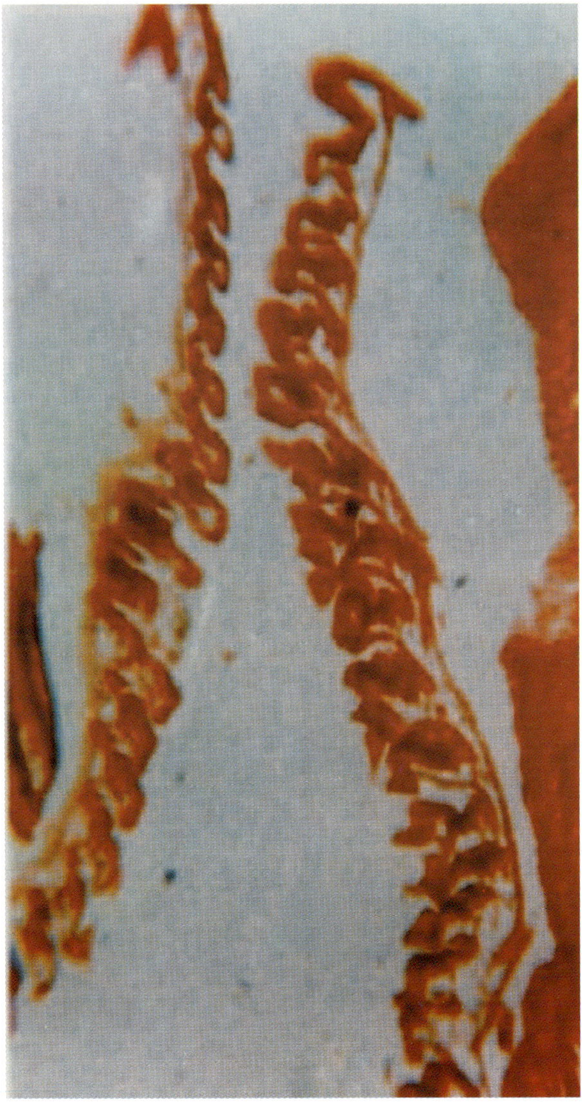

Pl. 27 E. Gill exposed to acute concentration of endosulfan. X 100.

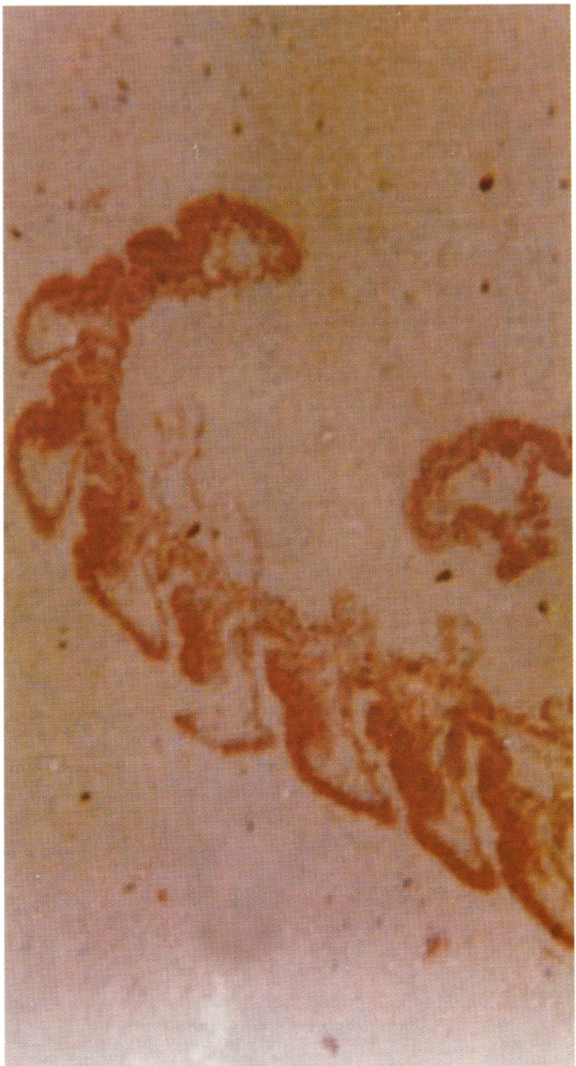

Pl. 27 F. Gill exposed to sublethal concentration of endosulfan. X 100.

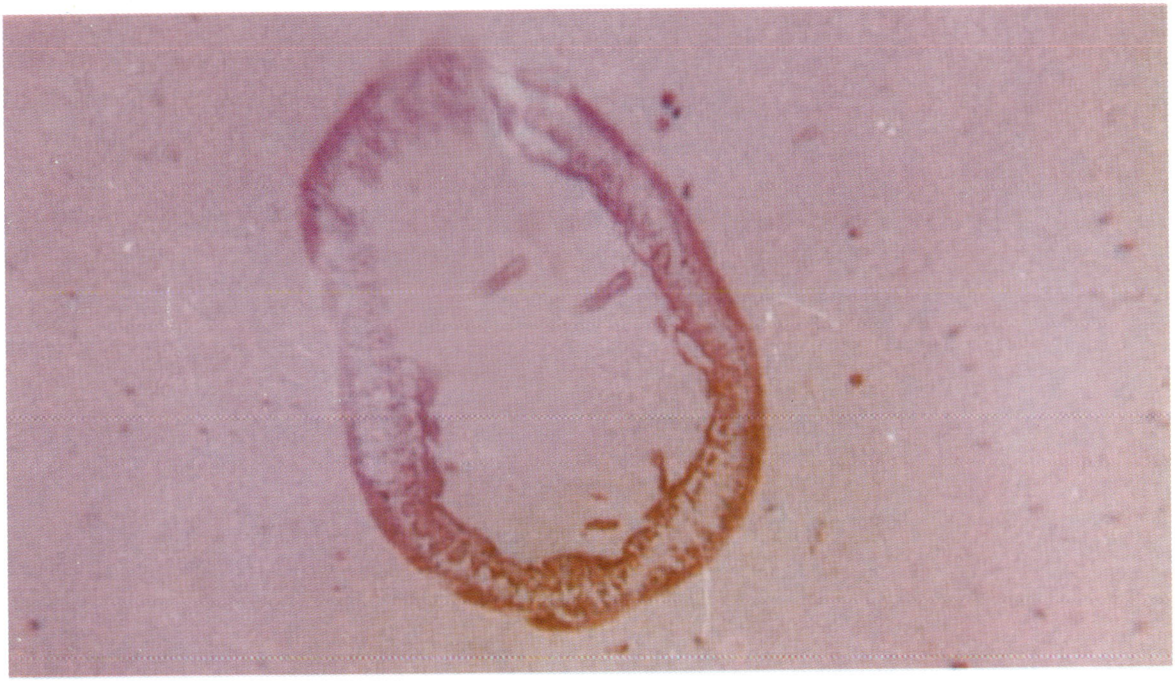

Pl. 29 E. C.S. of endusolfan exposed (acute) siphon of *T. ala.* X 40.

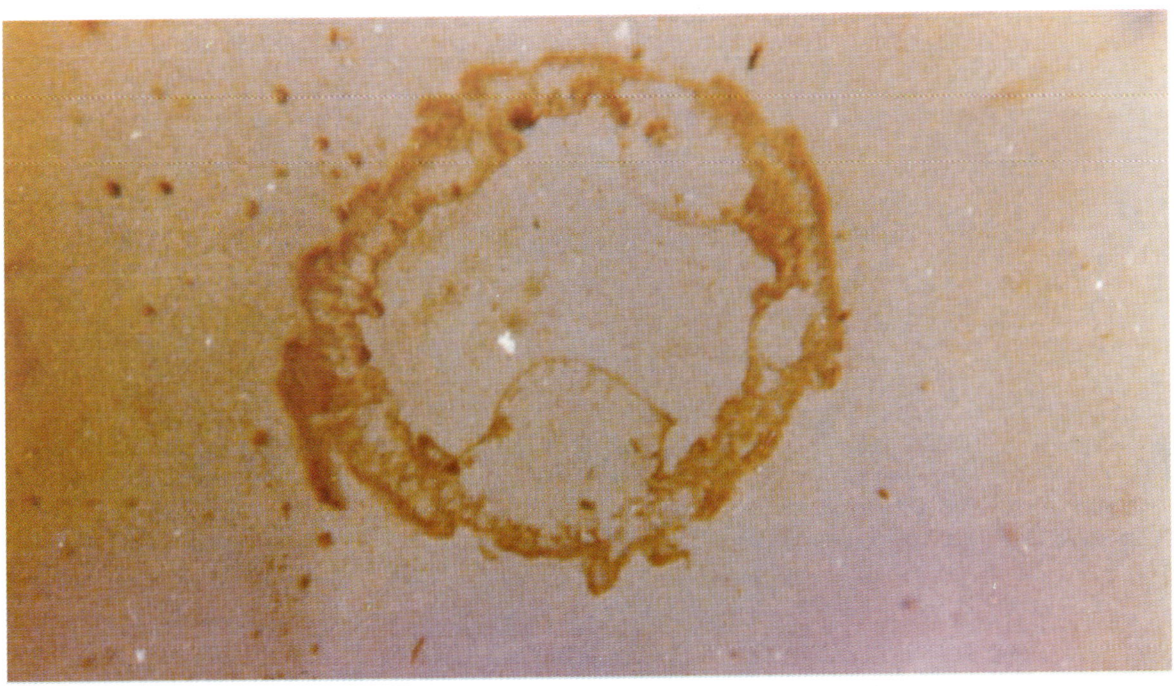

Pl. 29 F. C.S. of endusolfan exposed (sublethal) siphon of *T. ala.* showing dialation of siphonal wall. X 40.

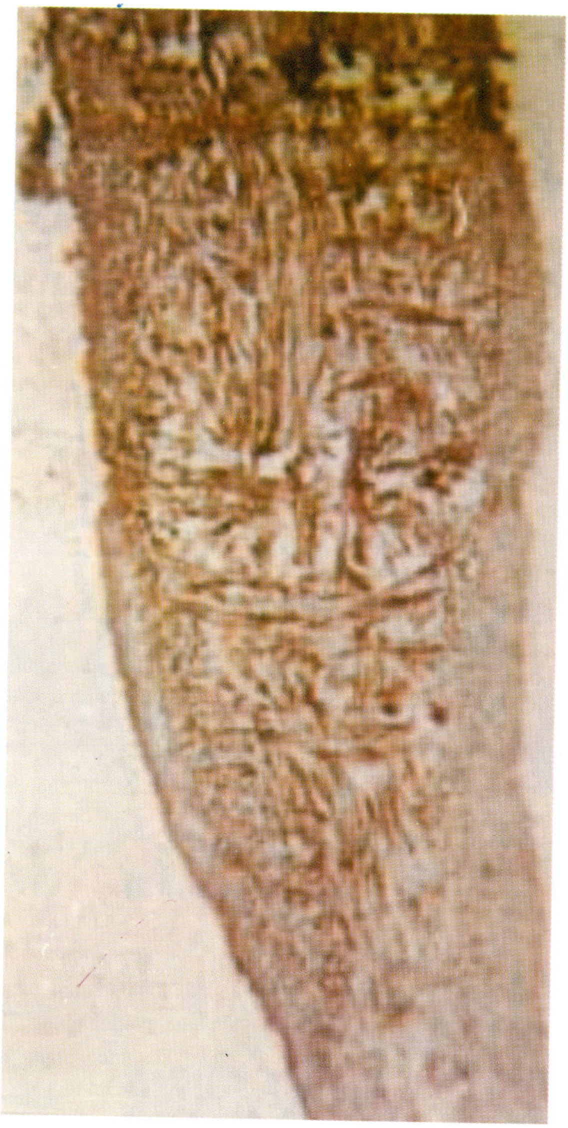

Pl. 30 A. T.S. of foot (control) of *T. ala.* X 100.

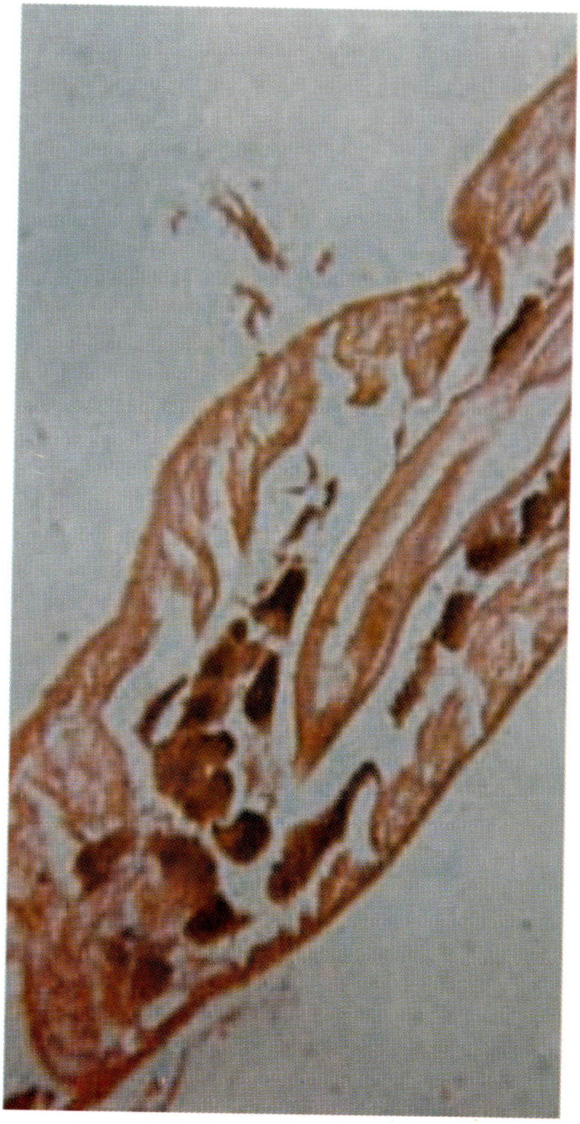

Pl. 30 B. T.S. of foot exposed to acute concentration of endusolfan. H and E. X 100.

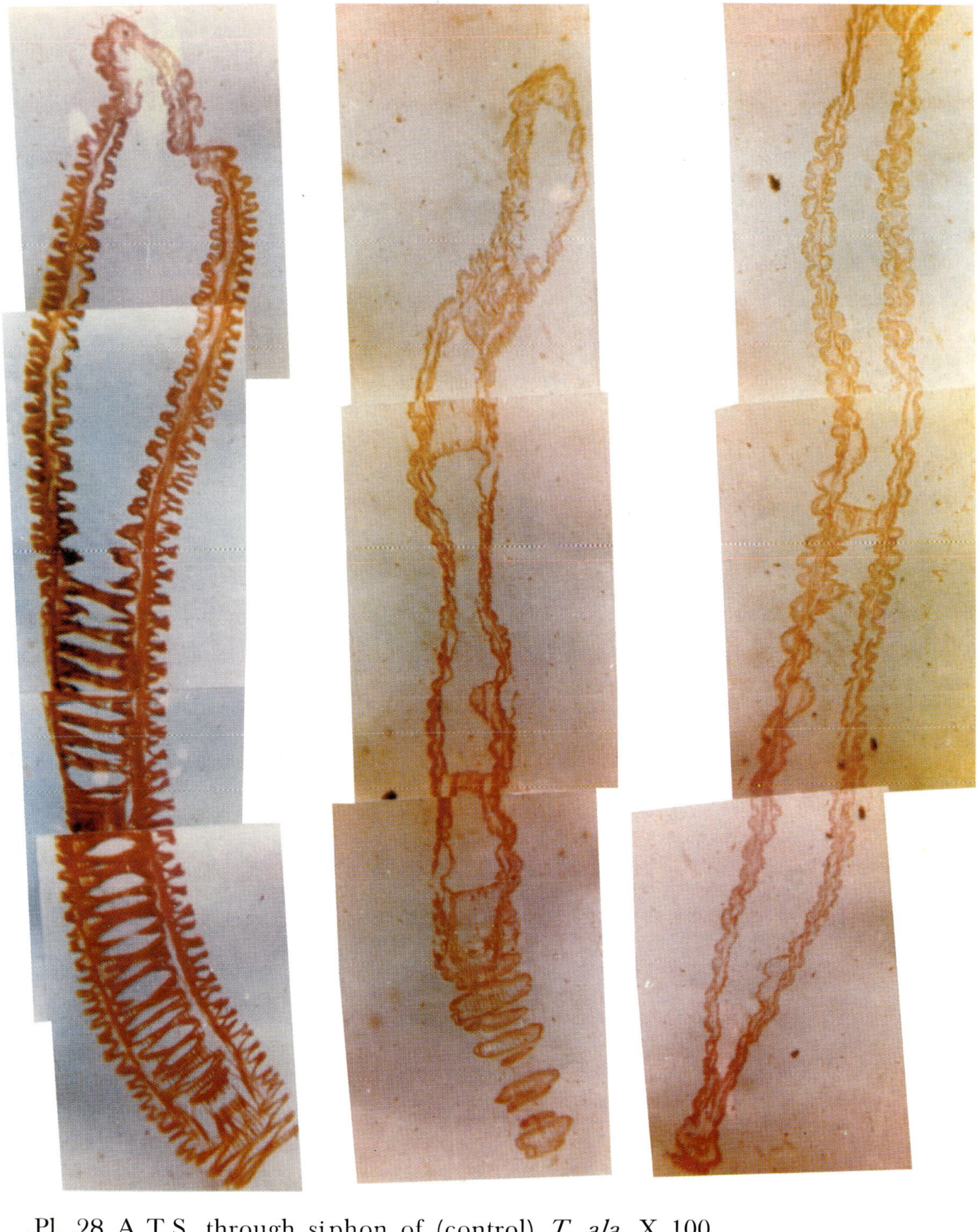

Pl. 28 A. T.S. through siphon of (control) *T. ala.* X 100.

Pl. 28 B. T.S. through siphon of *T. ala* exposed to acute concentration (10 ppb) of endosulfan. H and E. X 100.

Pl. 28 C. T.S. through sublethal concentration of endusolfan exposed siphon of *T. ala.* H and E. X 100.

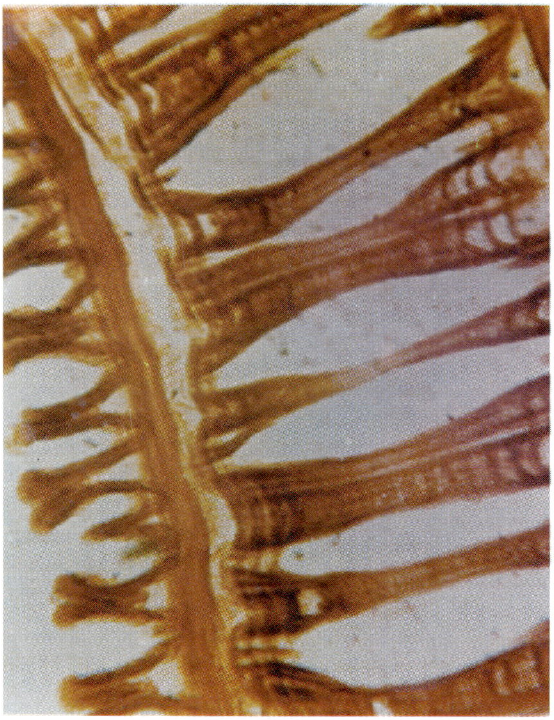

Pl. 29 A. A portion of control siphon of *T.ala*. H and E. X 400

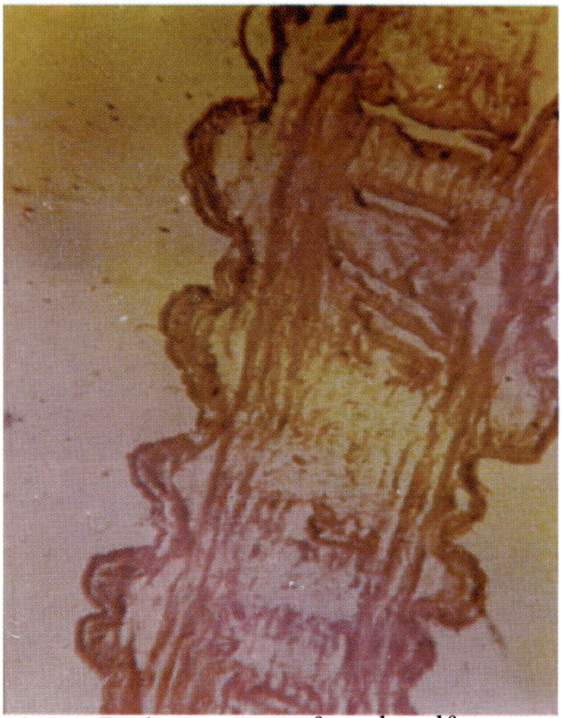

Pl. 29 B. A portion of endusolfan exposed (acute) siphon of *T. ala*. H and E. X 400.

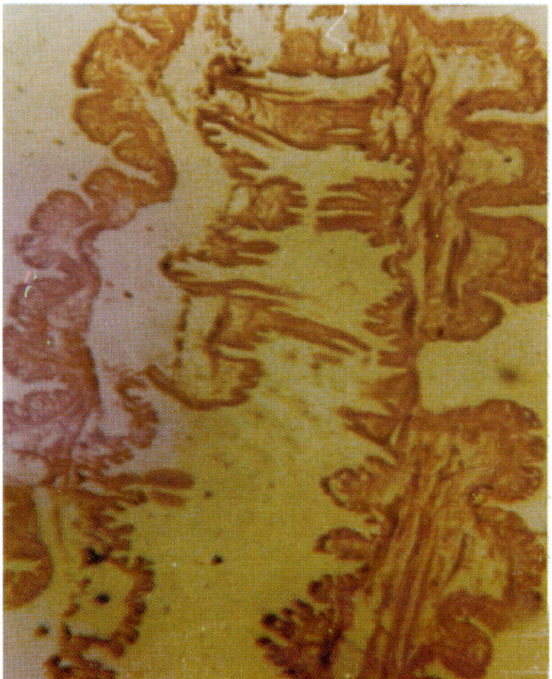

Pl. 29 C. A portion of endusolfan exposed (sublethal) siphon of *T. ala*. H and E. X 400.

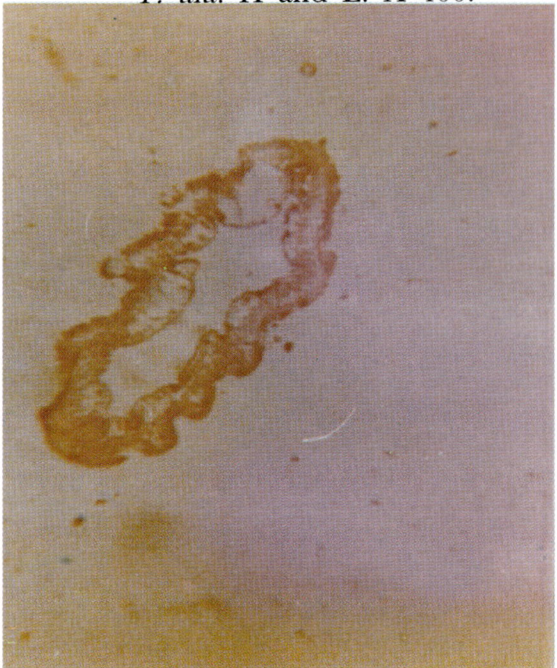

Pl. 29 D. C.S. of control siphon of *T. ala*. X 40.

Pl. 28 A. T.S. through siphon of (control) *T. ala*. X 100.

Pl. 28 B. T.S. through siphon of *T. ala* exposed to acute concentration (10 ppb) of endosulfan. H and E. X 100.

Pl. 28 C. T.S. through sublethal concentration of endusolfan exposed siphon of *T. ala*. H and E. X 100.

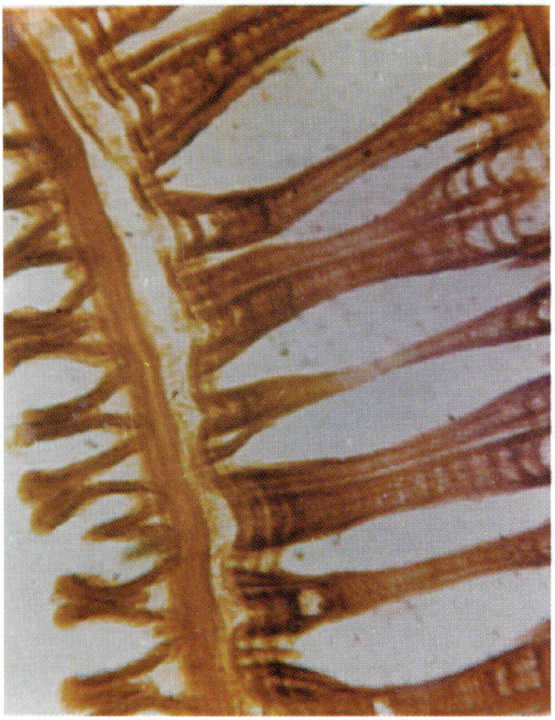

Pl. 29 A. A portion of control siphon of *T.ala.* H and E. X 400

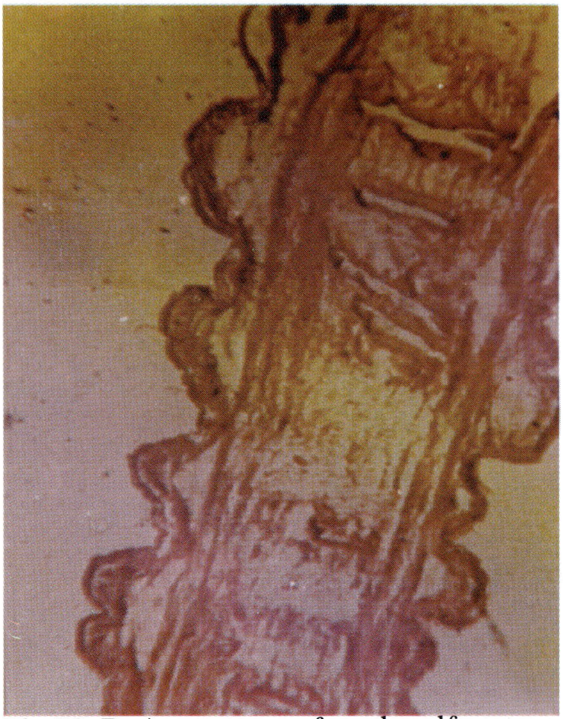

Pl. 29 B. A portion of endusolfan exposed (acute) siphon of *T. ala.* H and E. X 400.

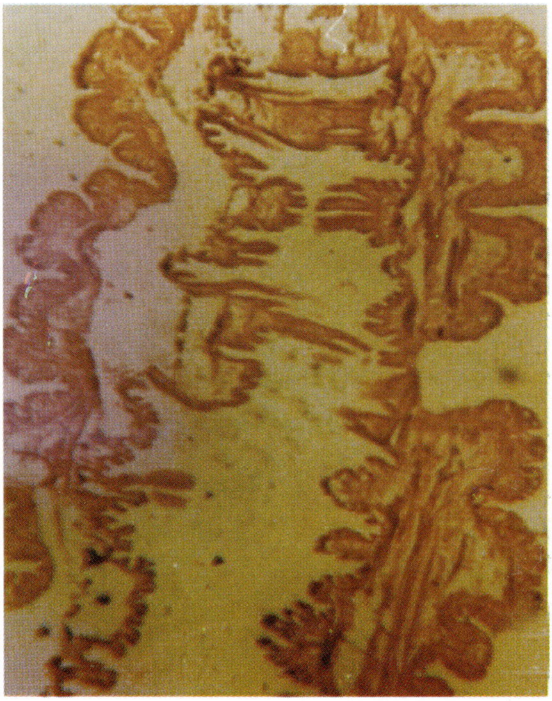

Pl. 29 C. A portion of endusolfan exposed (sublethal) siphon of *T. ala.* H and E. X 400.

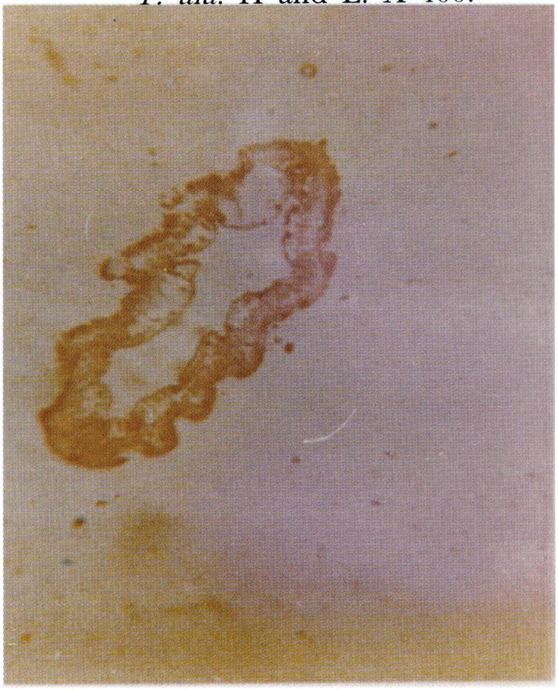

Pl. 29 D. C.S. of control siphon of *T. ala.* X 40.

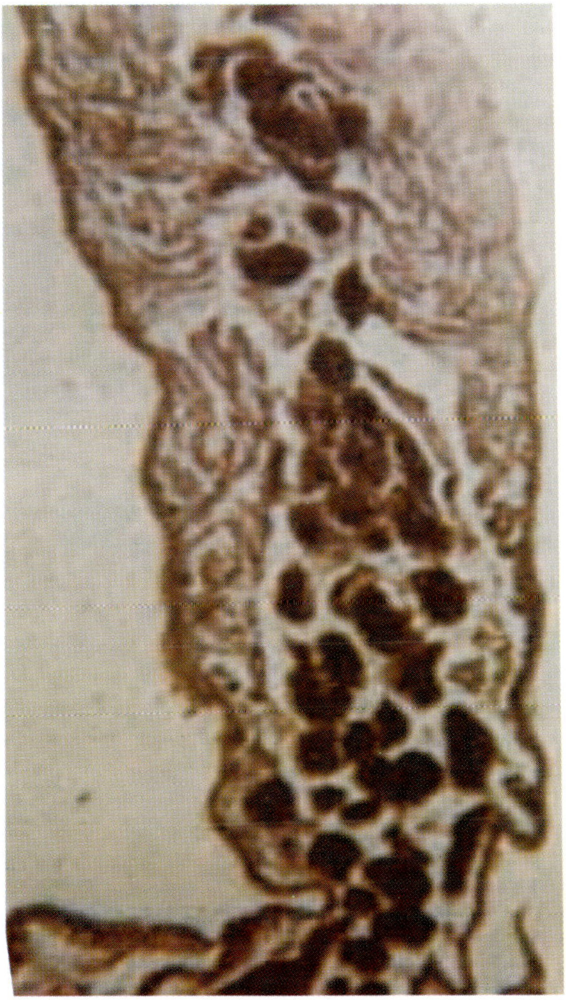

Pl. 30 C. T.S. of foot exposed to sublethal concentration of endusolfan. H and E. X 100.

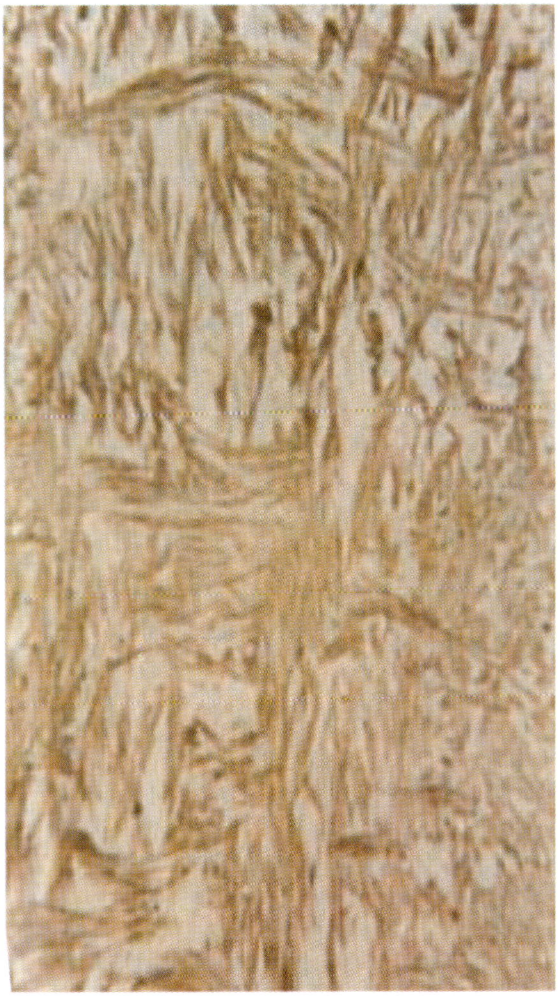

Pl. 30 D. A portion of control foot. H and E. X 400.

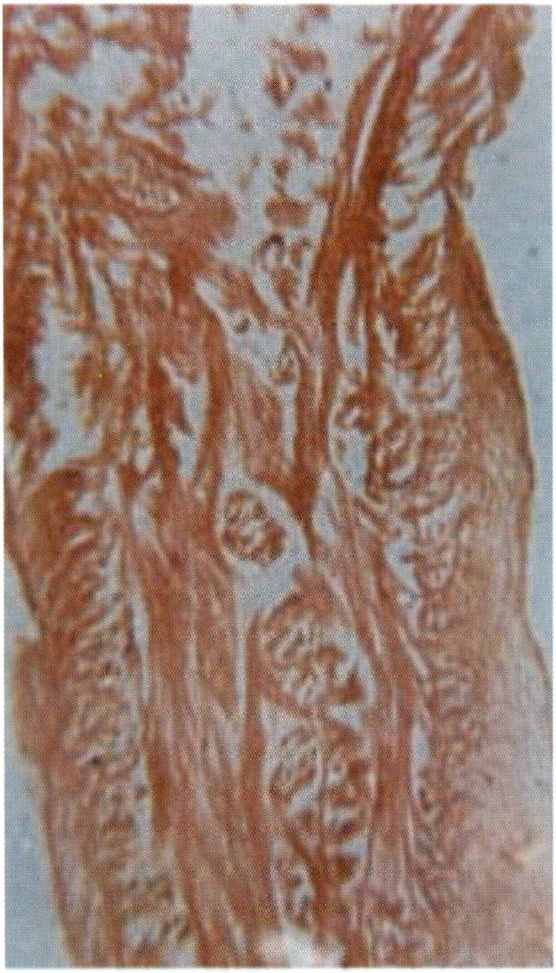

Pl. 30 E. A portion of endusolfan exposed (acute) foot. H and E. X 400.

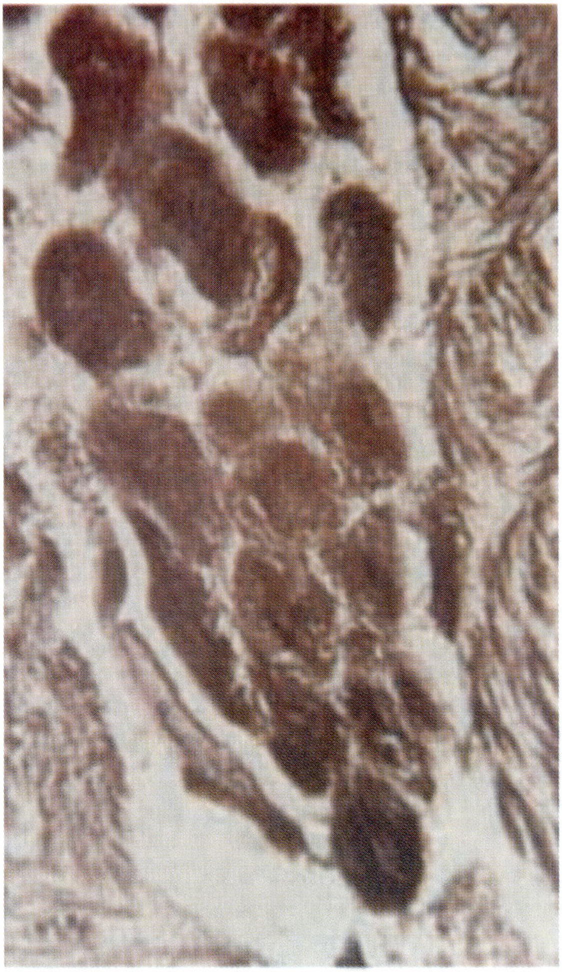

Pl. 30 F. A portion of endusolfan exposed (sublethal conc.) foot. H and E. X 400.

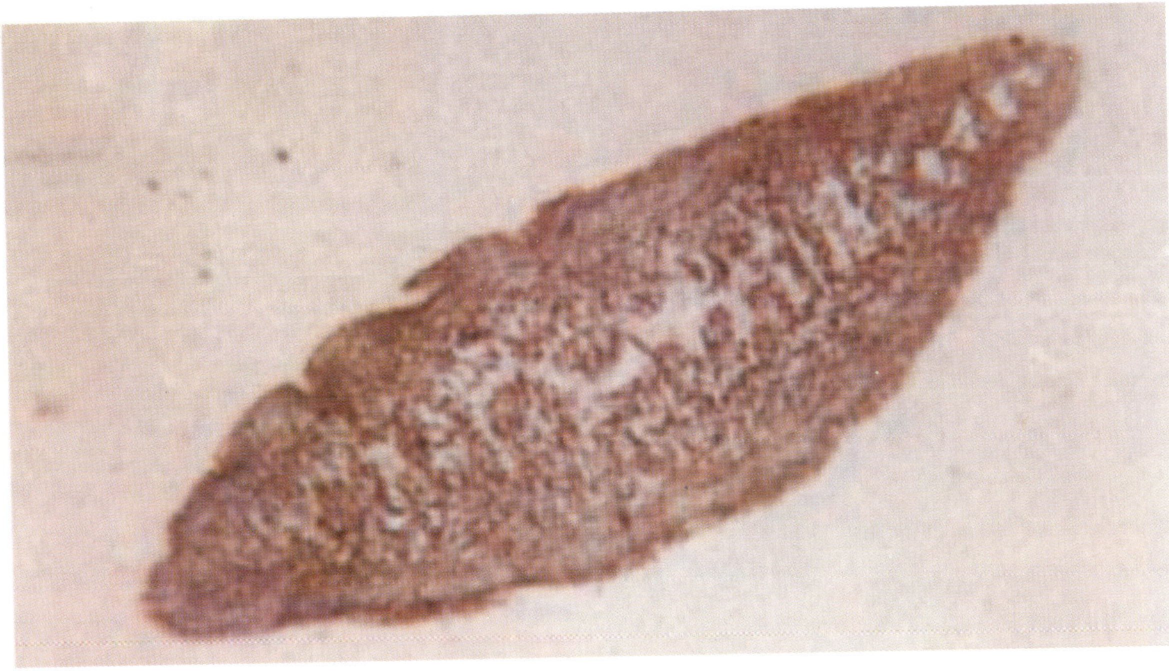

Pl. 30 G. Cross section of control foot of *T. ala* H and E. X 100.

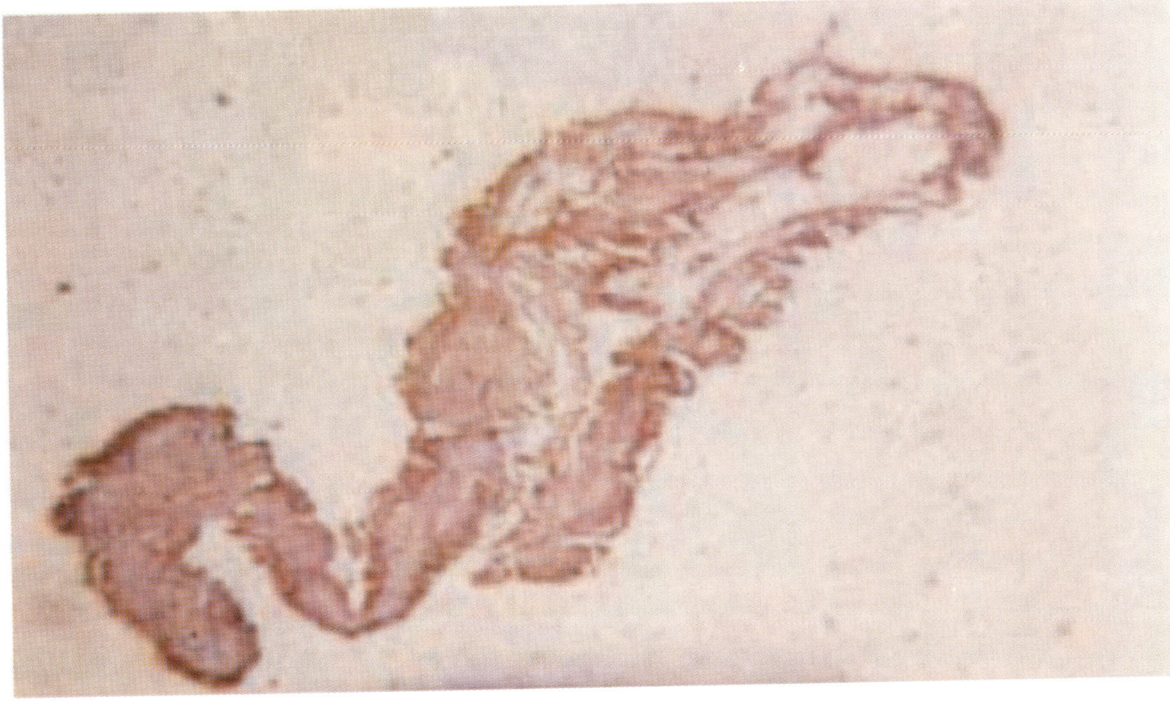

Pl. 30 H. Cross section of exposed (endusolfan acute conc.) foot to *T. ala* H and E. X 100.

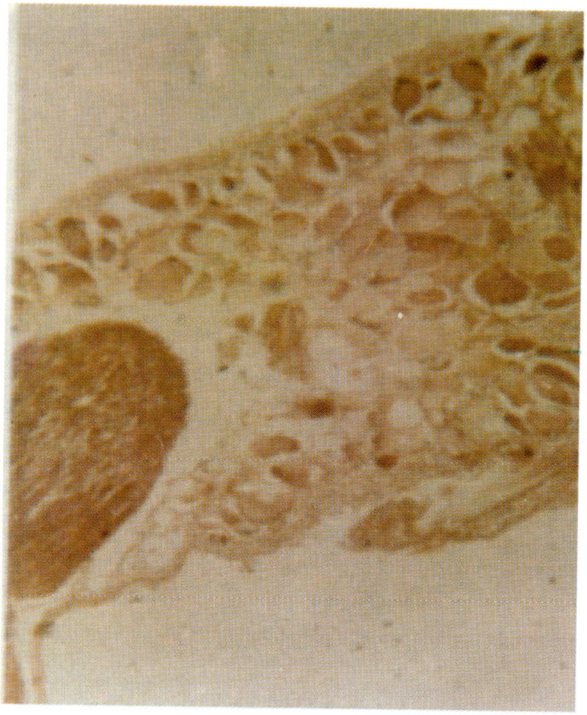

Pl. 31 A. Control gonad (oocyte) of *T. ala* H and E. X 100.

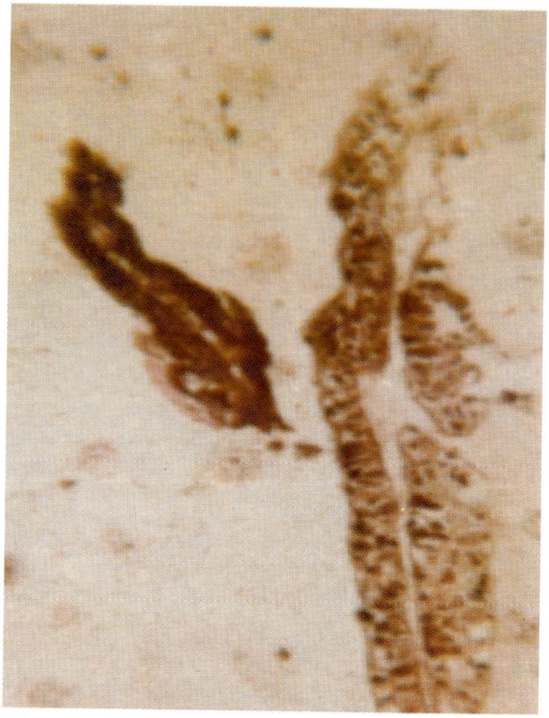

Pl. 31 B. Endusolfan exposed (acute conc.) gonad of *T. ala* H and E. X 100.

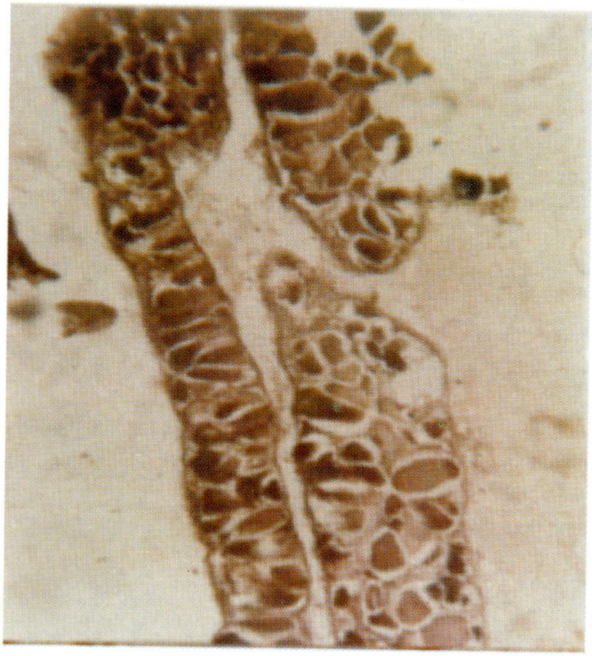

Pl. 31 C. Endusolfan exposed (acute conc.) gonad of *T. ala* H and E. X 400.

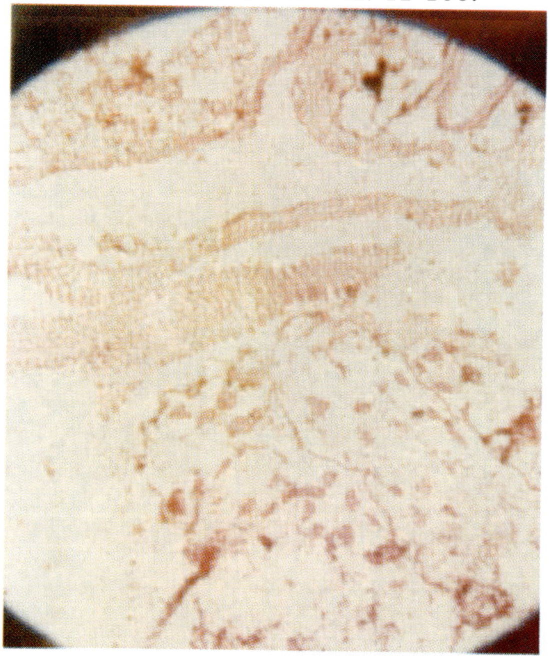

Pl. 31 D. Endusolfan exposed (sublethal concentration) gonad of *T. ala* H and E. X 30.

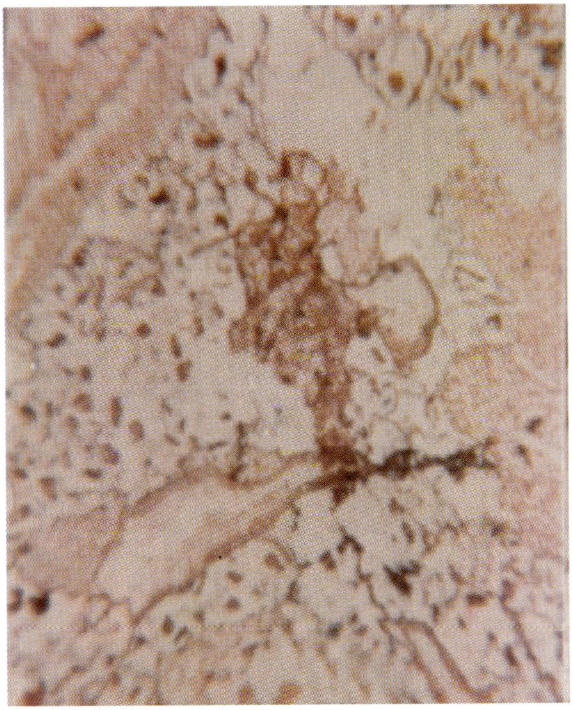

Pl. 31 E. Endusolfan exposed (sublethal concentration) gonad of *T. ala* H and E. X 100.

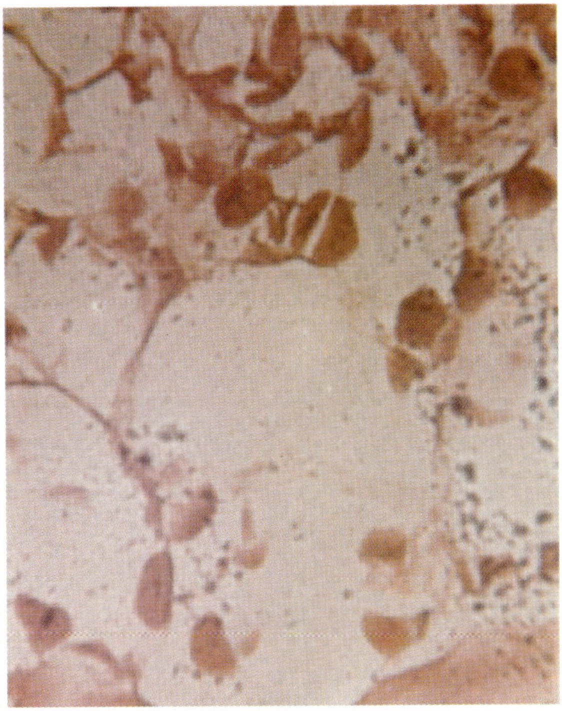

Pl. 31 F. Endusolfan exposed (sublethal concentration) gonad of *T. ala* H and E. X 400.

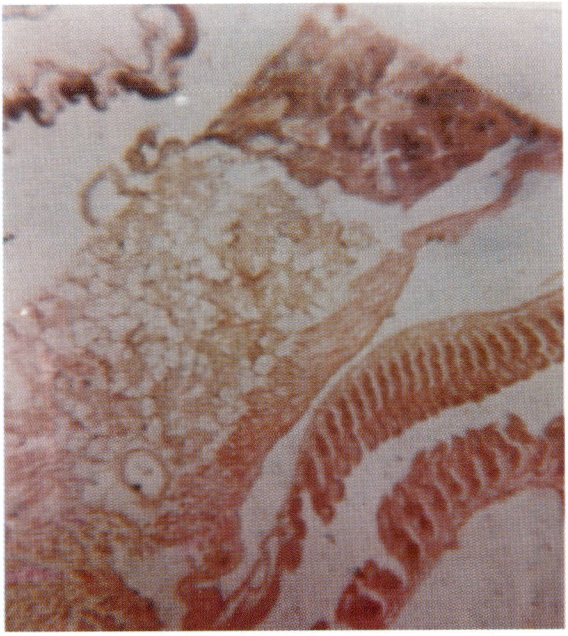

Pl. 32 A. Control digestive diverticula of *T. ala* H and E. X 30.

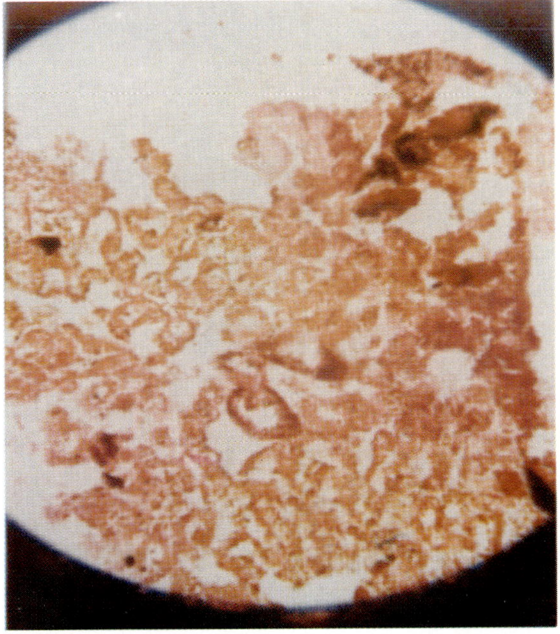

Pl. 32 B. Endusolfan exposed (acute) digestive diverticula of *T. ala* H and E. X 100.

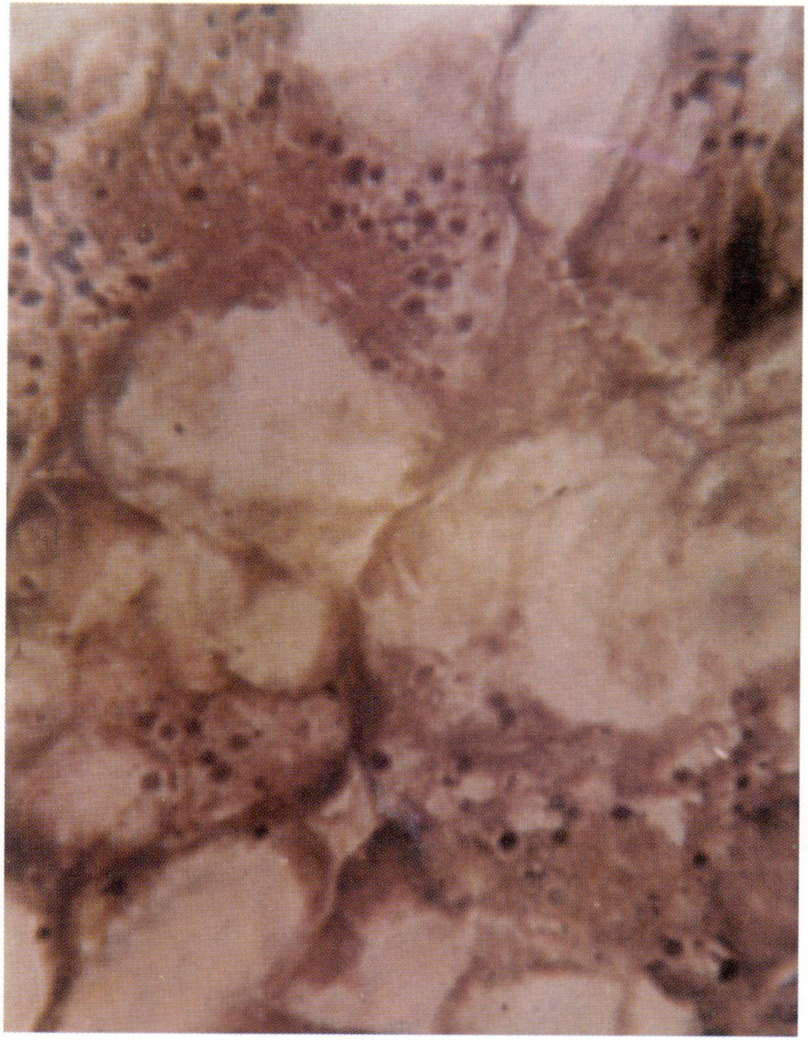

Pl. 32 C. Control digestive diverticula of *T. ala* H and E. X 400.

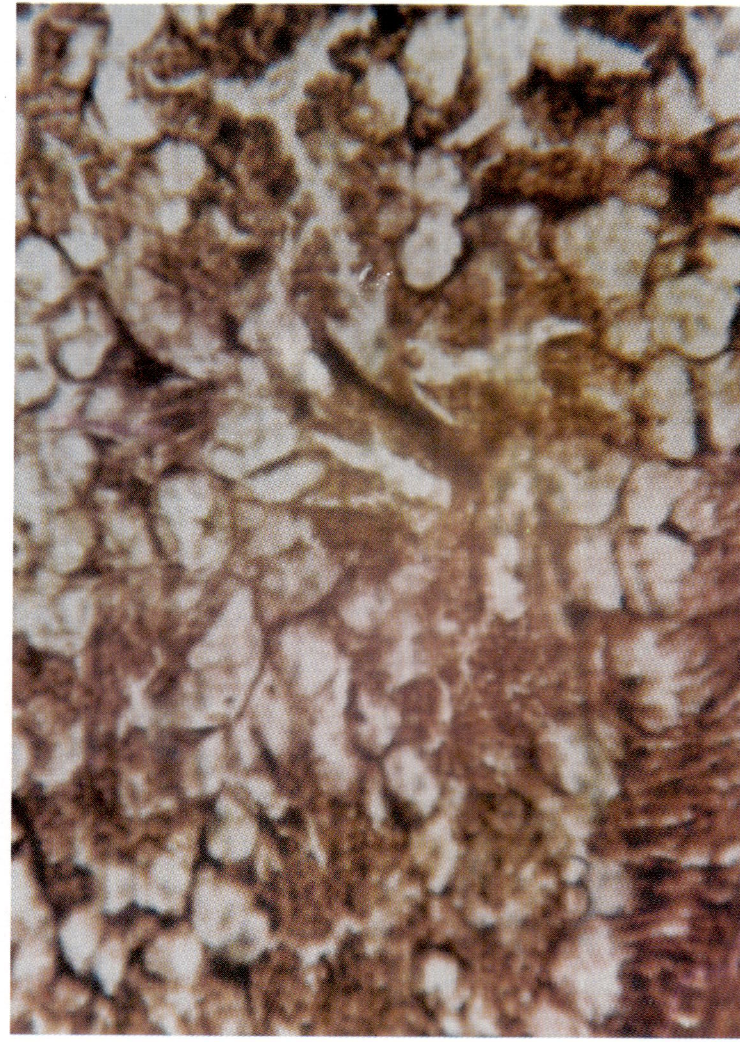

Pl. 32 D. Endusolfan exposed (sublethal concentration) digestive diverticula of *T. ala*. H and E. X 100.